Dissertation

The Construction of Nonseparable Wavelet Bi-Frames and Associated Approximation Schemes

Martin Ehler

2007

Bibliografische Information der Deutschen Nationalbibliothek

Die Deutsche Nationalbibliothek verzeichnet diese Publikation in der Deutschen Nationalbibliografie; detaillierte bibliografische Daten sind im Internet über http://dnb.d-nb.de abrufbar.

ISBN 978-3-8325-1771-7

Logos Verlag Berlin GmbH
Comeniushof, Gubener Str. 47,
10243 Berlin
Tel.: +49 030 42 85 10 90
Fax: +49 030 42 85 10 92
INTERNET: http://www.logos-verlag.de

The Construction of Nonseparable Wavelet Bi-Frames and Associated Approximation Schemes

Dissertation
zur
Erlangung des Doktorgrades
der Naturwissenschaften
(Dr. rer. nat.)

dem

Fachbereich Mathematik und Informatik
der Philipps-Universität Marburg

vorgelegt von

Martin Ehler
aus Frankenberg/Eder

Marburg/Lahn September 2007

Vom Fachbereich Mathematik und Informatik
der Philipps-Universität Marburg als Dissertation
angenommen am: 09. Oktober 2007

Erstgutachter: Prof. Dr. Stephan Dahlke, Philipps-Universität Marburg
Zweitgutachter: Prof. Dr. Gerlind Plonka-Hoch, Universität Duisburg-Essen
Drittgutachter: Prof. Dr. Manfred Tasche, Universität Rostock

Tag der mündlichen Prüfung: 19. Oktober 2007

Acknowledgements

First and foremost, I would like to thank Professor Stephan Dahlke, my thesis advisor, who endorsed and inspired me both to address the topic of wavelet frames and to consider new aspects while working on this thesis.

I am thankful to Professor Gerlind Plonka-Hoch for being my second referee and for organizing the Rhein-Ruhr Workshops, which are a pleasure. Many thanks to Professor Manfred Tasche for his willingness to write the third referee report.

Special thanks to Professor Wolfgang Gromes for lecturing on wavelets in 2001 and for his encouraging guidance as I began my studies in wavelet analysis.

Furthermore, I have to thank the members of the AG Numerik/Wavelet-Analysis and all of my other colleagues in Marburg for the kind work climate. An honorable mention goes to Thorsten, not only for his technical, but also for his general support.

I am much obliged to Anke Raufuß and Annie McWhertor Hamood for their straightforward help and their valuable comments and revisions. Special thanks go to Anke for keeping company up on the Lahnberge; it was great to share an office with you!

Many thanks to Daniel, not only for being a "soft skilled" colleague, but also for introducing me to my girlfriend, Sophie, and to David for helping me clear my mind with rounds of disc golf.

Last but not least, I thank my family for keeping me calm during my years of study and Sophie for her existence and for all that I cannot put into words.

Zusammenfassung

In nahezu allen technischen Anwendungen der heutigen Zeit müssen Daten analysiert und weiterverarbeitet werden. Solche Daten werden üblicherweise als Funktionen aufgefasst, deren Analyse ein Zerlegen in einfache Bausteine erfordert. In der Wavelet-Analyse werden die Bausteine durch Translatieren (Verschieben) und Dilatieren (Stauchen bzw. Strecken) endlich vieler Funktionen, die als Wavelets bezeichnet werden, erzeugt. Man kann Wavelets mit kompakten Trägern verwenden, so dass durch verschieden starkes Dilatieren feine oder grobe Auflösungen erreicht werden. Wir sprechen deshalb auch von einer Multiskalenauflösung, welche insbesondere zur Untersuchung lokaler Details einer Funktion notwendig ist. Dies stellt den wesentlichen Vorteil gegenüber der Fourier-Analyse dar, die Funktionen in ihre Frequenzanteile zerlegt. Deren Bausteine sind schlecht lokalisiert und auch die sogenannte gefensterte Fourier-Analyse lässt nur eine konstante Auflösung zu.

Die schnellen Algorithmen der Wavelet-Transformation werden bereits erfolgreich in der Signal- und Bildverarbeitung eingesetzt. Weitere Anwendungsgebiete sind Operator-Gleichungen, inverse Probleme und auch viele Arten von Variationsproblemen. Deren Lösung erfordert die Betrachtung spezieller Funktionenräume, im Wesentlichen sogenannte Besov-Räume. Vorteilhaft ist, dass diese durch orthonormale Wavelets charakterisiert werden, d.h. die Waveletsysteme bilden Basen in Besov-Räumen und die Norm des Besov-Raums kann durch eine äquivalente Folgennorm der Waveletkoeffizienten ausgedrückt werden. Damit stellen Wavelets eine effektive Diskretisierung des ursprünglichen Problems dar, was eine Grundvoraussetzung erfolgreicher Lösungsverfahren darstellt.

Das Zerlegen in einfache Bausteine erfordert auch wieder eine Rekonstruktion in Form einer Reihentwicklung. In der Praxis kann die Reihe nicht exakt berechnet werden. Deshalb versucht man, die Funktion durch eine möglichst gute Auswahl von N Bausteinen zu approximieren. Es ist wichtig, die zugehörigen Approximationsraten zu bestimmen. Für orthogonale Wavelet-Basen lassen sich diese Raten durch die Besov-Regularität der Wavelets und der jeweils zu approximierenden Funktion bestimmen.

Die oben genannten Anwendungen von Wavelets profitieren im Wesentlichen von inneren Waveleteigenschaften, z.B. kleinem Träger zur Lokalisation sowie Glattheit und verschwindende Momente für eine hohe Approximationsordnung. In vielen Anwendungen sind noch weitere Eigenschaften der Wavelets von Vorteil, vor allem die Symmetrie in der Signal- und Bildverabeitung.

Im Hinblick auf Konstruktionen betrachten wir zunächst univariate Wavelet-Basen. Orthogonale Wavelets wurden von Ingrid Daubechies erfolgreich und umfassend behandelt. Allerdings verhindert Orthogonalität wichtige zusätzliche Eigenschaften wie beispielsweise die Symmetrie. Um diesen Nachteil zu beseitigen, kann man zwei verschiedene Wavelet-Basen konstruieren, die biorthogonal zueinander stehen. Diese stellen weiterhin eine Reihenentwicklung ganz ähnlich zu orthogonalen Wavelets bereit, und sie erlauben symmetrische Wavelets.

In vielen Anwendungen benötigt man multivariate Wavelets. Während bei univariaten

Wavelets die Skalierung mit Zweierpotenzen kanonisch ist, führt diese dyadische Dilatation jedoch in höheren Dimensionen zu einem exponentiellen Anstieg der Anzahl benötigter Wavelets. Durch die dann steigende Komplexität werden die Wavelet-Algorithmen unbrauchbar. Um dies zu vermeiden, ersetzen wir den Faktor 2 durch eine sogenannte Dilatationsmatrix, d.h. durch eine ganzzahlige diagonalisierbare Matrix, deren sämtliche Eigenwerte einen Betrag größer eins haben. Man dilatiert dann anstelle der Zweierpotenzen mit den Potenzen der Matrix. Dies ermöglicht beispielsweise Wavelet-Basen in beliebigen Dimensionen, die nur aus einem einzigen Wavelet gebildet werden. Für isotrope Skalierungen, also diagonaliserbare Dilatationsmatrizen, deren Eigenwerte den gleichen Betrag haben, charakterisieren auch biorthogonale Wavelets noch Besov-Räume und die N-Term Approximationraten werden durch diese Räume bestimmt. Wir konzentrieren uns in der vorliegenden Arbeit auf dieser Form der Skalierung.

Konstruktionen von multivariaten biorthogonalen Wavelet-Basen leiden unter dem Nachteil, dass gute primale Wavelets in der Regel mit schlechteren dualen Wavelets gepaart werden müssen. Diesem Problem werden wir mit Hilfe des schwächeren Konzepts der Frames begegnen. Wavelet-Bi-Frames verallgemeinern Paare biorthogonaler Wavelet-Basen und bieten weiterhin eine stabile Zerlegung. Im Gegensatz zu Basen sind diese Systeme jedoch in der Regel redundant. Dieses Konzept bietet einen größeren Freiraum für Konstruktionsverfahren, den wir nutzen werden.

Mehrdimensionale Wavelet-Frame-Konstruktionen der bisher veröffentlichten Fachliteratur leiden entweder unter wenigen verschwindenen Momenten, fehlender Regularität oder einer zu großen Anzahl an Wavelets. In der vorliegenden Arbeit werden wir multivariate Wavelet-Bi-Frames konstruieren, die sich den Beschränkungen von Wavelet-Basen entziehen und bestehenden Framekonstruktionen überlegen sind. Allerdings müssen wir sicherstellen, dass wir die Charakterisierung von Funktionenräumen nicht verlieren und die N-Term-Approximationsraten noch bestimmt werden können.

Die obige Diskussion erfordert nunmehr die Lösung der folgenden vier Probleme:

(P1) Zeige, dass der Frameansatz genügend Flexibilität bietet, um die Beschränkungen von multivariaten Wavelet-Basen zu überwinden.

Wir versuchen optimale Resultate zu erzielen:

(P2) Stelle geeignete Optimalitätskriterien auf und konstruiere beliebig glatte Wavelet-Bi-Frames in beliebigen Dimensionen, die alle Optimalitätskriterien erfüllen.

Bisher konnte die Charakterisierung von Besov-Räumen und die Beschreibung der N-Term Approximation bezüglich Wavelet-Bi-Frames nur für dyadische Skalierungen gezeigt werden. Um die Anzahl der Wavelets zu minimieren, müssen wir jedoch allgemeinere Dilatationsmatrizen betrachten. Dazu benötigen wir eine Lösung des dritten Problems:

(P3) Charakterisiere Besov-Räume mittels Wavelet-Bi-Frames und beschreibe deren N-Term-Approximation auch für nichtdyadische Skalierungen.

Während Wavelet-Basen bereits erfolgreich in der Signal- und Bildverarbeitung Anwendung finden, müssen Wavelet-Frames noch zeigen, dass sie eine wertvolle Alternative darstellen können. Diese Forderung führt uns zum letzten Problem:

(P4) Weise die Nützlichkeit von Wavelet-Bi-Frames zu Anwendungszwecken nach. Demonstriere, dass Wavelet-Bi-Frames beim Entrauschen von Bildern gute Resultate liefern können.

Alle vier Probleme werden in der vorliegenden Arbeit gelöst. Die Resultate werden in der folgenden Inhaltsangabe vorgestellt.

Während wir im ersten Kapitel die Theorie der biorthogonalen Wavelet-Basen darstellen, führen wir in Kapitel 2 Wavelet-Bi-Frames ein und entwickeln die in (P2) erwähnten Optimalitätsbedingungen. Im dritten Kapitel konstruieren wir verschiedene Wavelet-Bi-Frames, die bis auf die Anzahl der Wavelets fast alle Optimalitätskriterien erfüllen. Schließlich erhalten wir unter anderem eine Familie beliebig glatter Wavelet-Bi-Frames in beliebigen Dimensionen mit nur drei Wavelets. Unsere konstruierten Wavelet-Bi-Frames sind im Vergleich zu biorthogonalen Wavelet-Basen glatter bei gleichzeitig höherer Approximationsordnung und kleinerem Träger. Damit wird (P1) gelöst.

In Kapitel 4 leiten wir eine Konstruktionsmethode her, die zu einer geringeren Anzahl an Wavelets führt. Neben weiteren Beispielen erhalten wir eine Familie beliebig glatter Wavelet-Bi-Frames in beliebigen Dimensionen mit nur zwei Wavelets, die alle Optimalitätskriterien erfüllen. Somit wird auch (P2) vollständig gelöst.

Für die Charakterisierung von Besov-Räumen mit Wavelet-Bi-Frames wiederholen wir in Kapitel 5 zunächst die bereits bekannten Resultate bezüglich biorthogonaler Wavelets mit isotroper Skalierung und dyadischen Wavelet Bi-Frames. Letztlich erweitern wir die Charakterisierung durch dyadische Wavelet-Bi-Frames auf isotrope Dilatationsmatrizen. Dies löst den ersten Teil von (P3).

Wir betrachten die N-Term-Approximation mit Wavelet-Bi-Frames in Kapitel 6. Um die Approximationsraten zu bestimmen, müssen wir sogenannte Jackson- und Bernstein-Ungleichungen herleiten. Dies gelingt zumindest für eine große Unterklasse von isotropen Skalierungen, was schließlich die Approximationsraten durch Besov-Räume bestimmt. Diese Beschränkung auf eine kleinere Klasse von Skalierungen stellt für uns de facto keine Einschränkung dar, weil alle Skalierungen der in den Kapiteln 3 und 4 konstruierten Wavelets dieser Unterklasse angehören. Abschließend zeigen wir, dass für die konstruierten Wavelets auch die weiteren Voraussetzungen der Jackson- und Bernstein-Ungleichungen erfüllt sind. Insofern lösen wir auch (P3) vollständig.

In Kapitel 7 entrauschen wir Bilder durch einen Variationsansatz in dem einer unserer in Kapitel 3 konstruierten Wavelet-Bi-Frames zur Diskretierung angewendet wird. Wir erhalten schließlich vielversprechende Resultate, die das Potential von Bi-Frames als sinnvolle Alternative zu Wavelet-Basen unterstreicht. Damit lösen wir (P4).

Contents

Introduction

Almost any kind of application requires at least to a certain extent the analysis of data. Depending on the specific application, the collection of data is usually called a measurement, a signal, or an image. In a mathematical framework, all of these objects are represented as functions. In order to analyze them, they are decomposed into simple building blocks. Such methods are not only used in mathematics, but also in physics, eletrical engeneering, seismic geology, wireless communication, target detection, and medical imaging.

In the nineteenth century, Fourier analysis was developed, where functions are decomposed into frequency components. However, these building blocks are very poorly localized, which causes serious problems in many applications. The windowed Fourier transform seemed to overcome this drawback by including a so-called window function. Yet, one often has to resolve a singularity of a given function. In other words, one has to be able to refine the resolution near a singularity. Unfortunately, once chosen, the window function is fixed, and one may speak of a constant resolution.

The development of wavelet theory is driven by the request for a more flexible tool and by the idea of variable resolution. In wavelet analysis, the building blocks are shifts and dilates of a finite number of functions $\psi^{(1)}, \ldots, \psi^{(n)}$, namely wavelets, i.e., one considers collections of the form

$$\left\{2^{\frac{jd}{2}}\psi^{(\mu)}(2^j x - k) : j \in \mathbb{Z},\ k \in \mathbb{Z}^d,\ \mu = 1, \ldots, n\right\}. \tag{0.1}$$

Wavelets with small supports provide a good localization as well as a flexible resolution according to different scaling indices $j \in \mathbb{Z}$, and many textbooks describe them as a mathematical microscope with which one can zoom in a function at a specific spot. Moreover, the fast wavelet transform provides the separation of signals into low- and high-frequency components, and it is nowadays successfully applied to signal and image processing to address compression, noise removal, and segmentation, see for instance the textbooks [Dau92, Mal99, SN96].

Many fields of applied mathematics, such as the numerical treatment of operator equations, inverse problems, and different kinds of variational methods, require the consideration of smoothness spaces. In order to derive solutions from practical algorithms, a discretization of the original problem is necessary. Wavelet analysis is a valuable tool beyond its fast transform since so-called Besov spaces, which cover most of the arising smoothness classes, are characterized by orthonormal wavelets, i.e., wavelets constitute bases for Besov spaces such that the original smoothness norm is equivalent to a weighted sequence norm of wavelet coefficients, cf. [DJP92]. Hence, decomposing into wavelet building blocks provides a discretization method, in which the continuous problem is replaced by a discret one in terms of wavelet coefficients.

Regarding the reconstruction, let us recall that bases provide series expansions. However, since algorithmic computations are limited to finite data, the series has to be replaced by a finite sum, let us say of length N. Then best N-term approximation is centered around the best choice of these terms, and it is essential to determine the approximation rate. Finally, in order to realize this rate in practical algorithms, one requires a simple rule for the choice of N terms. For orthonormal wavelet bases, it turns out that the approximation rate of a given function is determined by its Besov regularity. Simply taking the N largest coefficients of its series expansion provides a realization of the best N-term approximation rate, cf. [DJP92, Tem98]. Advantageously, the rule is simple and thresholding is computationally effective.

Before constructing wavelets, one has to identify which of their inner properties promote the above mentioned applications. Obviously, a very small support is essential for the idea of localization. Symmetric wavelets are claimed to provide better results in image and signal processing, cf. [Mal99], and many vanishing moments yield high compression rates. Moreover, smoothness and vanishing moments are ingredients for a high approximation order as well as for the characterization of function spaces. Finally, we identified important inner properties of wavelets such as

- small support,
- symmetry,
- smoothness,
- a large number of vanishing moments.

Unfortunately, smoothness and support sizes are competing properties, and constructions have to provide a certain balance between the two. The early ad-hoc wavelet constructions by Mallat, Meyer, and Stromberg exemplify unbalanced wavelet properties, cf. [Haa10, Mey86, Str81]. Mallat and Meyer then proposed a systematical method by introducing the concept of a so-called multiresolution analysis. It provides a powerful framework since the wavelet construction is essentially reduced to the construction of a refinable function φ, i.e, there is a coefficient sequence $(a_k)_{k\in\mathbb{Z}^d}$, namely the mask, such that

$$\varphi(x) = \sum_{k\in\mathbb{Z}^d} a_k \varphi(2x-k), \tag{0.2}$$

see [Mal89, Mey90] for details. Orthonormal wavelet bases, which seem most desirable according to Parseval's Equality, require that φ has orthonormal integer shifts. Nowadays, the multiresolution analysis framework is a standard tool for the construction of wavelets, and it also provides the fast wavelet transform.

Regarding the transform, there arises a further desirable property concerning the underlying refinable function of the wavelet basis. Given a function f to be analyzed, the exact determination of the input sequence for the transform is generally complicated and computationally expensive, cf. [Mal99, SN96]. Nevertheless, if the underlying refinable function φ is

- fundamental,

which means it is continuous and its shifts interpolate the integer grid, i.e.,

$$\varphi(k) = \delta_{0,k}, \quad \text{for all } k \in \mathbb{Z}^d,$$

then the input is determined as a sequence of sample values of f, see Subsection 1.2.2 for details. Thus, fundamental refinable functions simplify the exact application of the wavelet transform.

Returning to the construction of wavelets, we shall begin with the univariate setting. By applying the multiresolution analysis framework, Daubechies could construct her famous family of arbitrarily smooth compactly supported orthogonal wavelet bases, cf. [Dau92]. However, for many applications, we need multivariate wavelets and the aforementioned approach cannot be adapted to this multivariate setting since it uses a factorization technique of trigonometric polynomials, which does not hold in higher dimensions. Hence, a different method is required, and one very often uses tensor products of univariate wavelets. Unfortunately, such bases prefer the axis directions, and this is inconvenient for the visual perception of processed images, cf. [Mal99]. Moreover, tensor wavelet bases consist of 2^d-1 wavelets, which causes problems in higher dimensions since the complexity of the transform increases exponentially. In order to reduce the number of wavelets, we consider a different notion of scaling throughout the present thesis, i.e., we replace the dyadic dilation factor 2 in (0.1) and (0.2) by a so-called dilation matrix M, i.e., an integer matrix whose eigenvalues are larger than one in modulus. This concept allows for a finer scaling, and the number of required wavelets equals $m-1$, where $m = |\det(M)|$, which is independent of the dimension, cf. [CD93]. Then a dilation matrix with $m = 2$ allows for bases with only one wavelet, and such a bivariate choice is the popular quincunx matrix

$$M_q = \begin{pmatrix} 1 & -1 \\ 1 & 1 \end{pmatrix}. \tag{0.3}$$

However, there is still a lack of promising construction methods for smooth multivariate orthonormal wavelets with small supports. For instance, compactly supported and one times differentiable orthonormal wavelets for the quincunx dilation matrix in (0.3) are completely unknown so far. To make matters even worse, a compactly supported orthonormal wavelet basis with respect to a dilation matrix with $m = 2$ neither allows for symmetries nor for an underlying fundamental refinable function, cf. [Dau92, Han04] as well as Lemma 1.3.1 in the present thesis. One circumvents such problems with the concept of pairs of biorthogonal wavelet bases, i.e., one has primal and dual wavelets $\psi^{(\mu)}$ and $\widetilde{\psi}^{(\mu)}$, $\mu = 1, \dots, m-1$, respectively, whose dilates and shifts constitute two bases, which are biorthogonal to each other. This concept provides a series expansion similar to orthonormal bases, i.e.,

$$f(x) = \sum_{\mu=1}^{m-1} \sum_{j\in\mathbb{Z}} \sum_{k\in\mathbb{Z}^d} m^j \left\langle f, \widetilde{\psi}^{(\mu)}(M^j \cdot -k)\right\rangle \psi^{(\mu)}(M^j x - k), \tag{0.4}$$

and their construction is reducible to a pair of biorthogonal refinable functions φ and $\widetilde{\varphi}$, i.e.,

$$\langle \varphi(\cdot + k), \widetilde{\varphi}(\cdot + l)\rangle = \delta_{k,l}.$$

At least for isotropic scalings, i.e., dilation matrices that are diagonalizable and whose eigenvalues have the same modulus, biorthogonal wavelets still provide the characterization of Besov spaces as well as N-term approximation similar to dyadic orthonormal bases, cf. [Lin05]. It turns out that biorthogonal wavelets allow for symmetries, and the primal refinable function can be fundamental, see [CHR00, DGM99, DM97, Der99, HJ98,

HJ02, HR02, JRS99]. However, these constructions still bear some limitations. One can obtain strong properties of the primal wavelets, such as smoothness, small support, and a fundamental underlying refinable function. Unfortunately, these strong properties are generally accompanied with weak dual properties, i.e., dual wavelets have either poor smoothness or large support.

In order to overcome the limitations, one may proceed in two different directions. In [Koc07, Koc], one circumvents the aforementioned restrictions at least to a certain extent in the bivariate setting by switching to refinable vectors. However, this vector approach provides more complex structures in the fast wavelet transform, and the transform requires the conversion of the original signal into a vector structure, which seems a bit artificial and often causes computational problems.

In the present thesis, we avoid the vector setting, and, thus, follow a different approach. We attempt to circumvent the restrictions of orthogonal and biorthogonal wavelet bases with the weaker concept of frames. They still allow for a stable decomposition, and so-called wavelet bi-frames provide a series expansion very similar to those of pairs of biorthogonal wavelet bases in (0.4). Contrary to biorthogonal bases, primal and dual wavelets are no longer supposed to satisfy any geometrical conditions, and the frame setting allows for redundancy, i.e., there might possibly be coefficients different from inner products such that a series expansion converges towards the same function. The coupling of primal and dual wavelets in a bi-frame is much weaker than in the bases setting, and this yields more flexibility in their construction. For instance, smooth, symmetric wavelet frames with small support and a high number of vanishing moments have successfully been constructed in [CHS02, DHRS03, SA04], see [RS97b, RS97c] for some background information. However, these constructions are restricted to the univariate setting, and they apply certain factorization techniques, which do not hold in higher dimensions.

So far, all multivariate wavelet frame constructions in the literature suffer from the absence of desirable wavelet properties. We are unable to present a comprehensive list of multivariate constructions, but we shall point out a few. For instance, the approaches in [GR98, LS, RS98] suffer from a lack of vanishing moments. In [LS], one also derives wavelets with a high number of vanishing moments, but it is paid for by the loss of compact support. Smooth dyadic wavelet frames with a high number of vanishing moments are obtained from bivariate box splines in [CH01], but the method leads to a large number of wavelets. The general construction given in [Han03a] considers neither symmetry nor any optimality constraints.

The limitations of orthonormal, biorthogonal, and existing frame constructions are the motivation of the present thesis. By avoiding the vector approach, we attempt to construct multivariate wavelet frames, which circumvent the restrictions of the bases setting and which overcome the limitations of existing frame constructions. Finally, we have to find multivariate wavelet bi-frames that possess superior properties in terms of support sizes, smoothness, and vanishing moments, while providing a fast transform, the characterization of function spaces, and a description of N-term approximation rates.

Four problems result from the above discussion. The first addresses the potential of bi-frames in comparison to biorthogonal wavelet bases:

(P1) Verify that the frame approach provides sufficient flexibility to overcome the restrictions of multivariate wavelet bases. Construct multivariate wavelet bi-frames that inherit much better properties than biorthogonal wavelet bases.

In the second problem, we have to consider wavelet bi-frame constructions within the bi-frame setting:

(P2) Establish certain reasonable optimality criteria, and find optimal wavelet bi-frames. Moreover, construct families of arbitrarily smooth wavelet bi-frames in arbitrary dimensions that satisfy all optimality conditions.

As mentioned above, orthonormal and biorthogonal wavelet bases with general isotropic dilation matrices characterize Besov spaces, and their N-term approximation is well understood. This powerful framework must not be given up in the weaker frame setting. Borup, Gribonval, and Nielsen could derive an extension to wavelet bi-frames with dyadic scaling, cf. [BGN04]. Then norm equivalences hold with respect to the bi-frame coefficients, the best N-term approximation rate is determined by the Besov regularity, and the rate can be realized by thresholding the wavelet bi-frame expansion. Note that it may fail if one considers arbitrary expansions, in which the coefficients are not derived from inner products with the dual wavelets. Fortunately, the last mentioned particularity of the weaker frame setting accounts for few limitations in applications. Since the bi-frame results only address dyadic dilation, we have to consider the following third problem:

(P3) Characterize Besov spaces by wavelet bi-frames with general isotropic scalings, and extend the dyadic results about N-term approximation with wavelet bi-frames.

The final problem addresses the usefullness of wavelet bi-frames for applicational purposes. Since wavelet bases have already been successfully applied to different kinds of noise removal, we have to verify that wavelet bi-frames may constitute a valuable alternative. In [CDLL98], Chambolle, DeVore, Lee, and Lucier use orthogonal and biorthogonal wavelet bases for variational image denoising, i.e., they consider a variational problem with respect to Besov spaces depending on a so-called regularization parameter, which determines the amount of noise removal. By applying the characterization of Besov spaces, they derive an equivalent discrete variational problem in terms of wavelet coefficients. Then an approximate solution of the original problem can be derived by thresholding the wavelet coefficients, yet one still needs a method for choosing of an adequate regularization parameter. The H-curve criterion in [MP03] is a possible candidate. However, it has only been applied to discretizations by orthonormal wavelet bases so far. Finally, the application of wavelet bi-frames requires the solution of the following fourth problem:

(P4) In order to verify the usefulness of wavelet bi-frames for image denoising via a variational approach, establish the discretization of variational problems with respect to wavelet bi-frames, and demonstrate that the H-curve criterion provides decent results for bi-frames as well.

In the present thesis, we solve (P1), (P2), (P3), and (P4), as we shall explain in the following outline, where the problems are revisited.

Layout

The present thesis is organized as follows: In Chapter 1, we present an overview of multivariate biorthogonal wavelet bases with general dilation matrices. In order to address their construction, we recall the multiresolution analysis framework. Then, we consider desirable properties of wavelet systems in detail, and we place them on a wish list. Referring to the list, we discuss restrictions of wavelet bases concerning (P1). In order to circumvent these limitations, we address the weaker concept of wavelet frames in Chapter 2. We introduce wavelet bi-frames, and we recall a general framework for their construction, namely the mixed extension principle as proposed in [CHS02, DHRS03], where wavelets are still derived from an underlying refinable function. Finally, we discuss desirable wavelet bi-frame properties, and we establish optimality criteria concerning (P2). These criteria include the following: first, given a certain mask support, we derive statements about the maximal smoothness and the maximal approximation order offered by the underlying refinable function. Next, we address the approximation order of the wavelet bi-frame, and it turns out that the order is optimal if the wavelets have a sufficient number of vanishing moments. It should be mentioned that we do not consider N-term approximation in this chapter, but approximation with respect to a truncation of the bi-frame expansion. Finally, we address the maximal symmetry of wavelet bi-frames and the minimal number of wavelets provided that the underlying refinable function is fundamental.

The conceptual restrictions of biorthogonal wavelet bases are circumvented in Chapter 3 by the construction of frames. We derive smooth multivariate wavelet bi-frames for general scalings with small support satisfying a variety of extra conditions, such as symmetry and a large number of vanishing moments. The number of wavelets depends only on the dilation matrix, and in order to minimize that number, we choose a matrix with a determinant equal to ± 2. Then we obtain bi-frames with only three wavelets. Moreover, primal and dual wavelets are obtained from one single refinable function, which is even fundamental. This is impossible within the concept of biorthogonal wavelets. In the bivariate setting, we construct a family of arbitrarily smooth wavelet bi-frames for the popular quincunx dilation matrix in (0.3). We also obtain a dyadic bi-frame with the underlying box spline refinable function derived in [RS97a]. For specific dilation matrices satisfying $|\det(M)| = 2$, we construct a family of arbitrarily smooth wavelet bi-frames in arbitrary dimensions with three wavelets. Finally, all of our bi-frames provide significantly smaller supports in comparison to biorthogonal approaches, and they satisfy many optimality criteria established in Chapter 2. Hence, we solve problem (P1) completely and (P2) at least to a certain extent. The results presented in this chapter have been published in [Ehl].

In Chapter 4, we derive a wavelet bi-frame construction with fewer wavelets. Contrary to the previous chapter, we apply the mixed oblique extension principle as derived in [CHS02, DHRS03], see also [DH00, Han03b], which generalizes the mixed extension principle. As far as we know, we present its first multivariate application yielding compactly supported wavelets. Then we obtain wavelet bi-frames, whose underlying refinable functions have already been addressed in Chapter 3, but we reduce the number of wavelets. In particular, we obtain a family of arbitrarily smooth wavelet bi-frames in arbitrary dimensions with only two wavelets satisfying all of the optimality conditions established in Chapter 2. Hence, we solve (P2) completely. The results of this chapter have been published in [Ehl07].

The remaining chapters are dedicated to the problems (P3) and (P4). In Chapter 5, we introduce Besov spaces in detail. Then, we recall their characterization by pairs of biorthogonal wavelet bases with general isotropic scalings. In order to derive the wavelet bi-frame characterization with respect to those scalings, we try to follow the dyadic ideas of Borup, Gribonval, and Nielsen in [BGN04]. In a sense, they localize the dyadic bi-frame to a dyadic orthonormal wavelet basis, which plays the role of a reference system, such that the orthonormal characterization carries over to the bi-frame. However, in order to consider general isotropic scalings, a conceptual difficulty arises, because, for many dilation matrices as for instance the quincunx matrix in (0.3), sufficiently smooth orthonormal wavelets with compact support are not known. Hence, we require a different reference system. Fortunately, for many dilation matrices, we can find smooth compactly supported biorthogonal wavelets that provide the characterization of Besov spaces, cf. [Der99, JRS99]. Then, we generalize the localization technique regarding general isotropic dilation and biorthogonal reference systems. Finally, the biorthogonal characterization carries over to the wavelet bi-frame. This yields the solution to the first part of problem (P3).

In numerical analysis and approximation theory, one has to establish so-called matching Jackson and Bernstein inequalities in order to describe best N-term approximation. We derive both estimates with respect to wavelet bi-frames in Chapter 6. With the norm equivalences of Chapter 5 in hand, we can follow the approach in [BGN04] to obtain the Jackson inequality for general isotropic scalings. The required Bernstein estimate can be reduced to a Bernstein inequality involving only the underlying refinable function. In [Jia93], such an inequality is derived for dyadic dilation $M = 2\mathcal{I}_d$. An analysis of its proof reveals that it still holds for dilation matrices of the form $M = h\mathcal{I}_d$, $h \in \mathbb{N}$. Then, we establish an extension to idempotent scalings, i.e., dilation matrices M, which satisfy $M^l = h\mathcal{I}_d$, for some $l, h \in \mathbb{N}$. Fortunately, most isotropic dilation matrices in the literature are idempotent. For instance, the quincunx matrix in (0.3) satisfies $M^8 = 16\mathcal{I}_2$. In conclusion, at least for idempotent scalings, we establish matching Jackson and Bernstein estimates, which provide the description of best N-term approximation.

Finally, we address the realization of the approximation rate. It turns out that the rate can be realized by thresholding since the associated result regarding unconditional bases in [BN] only requires those properties that wavelet bi-frames also inherit. This provides the final solution to problem (P3).

In the remainder of Chapter 6, we verify that the wavelet bi-frames from Chapters 3 and 4 satisfy the requirements for the Jackson and Bernstein inequalities. On the one hand, this completes our construction of wavelet bi-frames since we describe their associated N-term approximation. On the other hand, it ensures that the theoretical results about N-term approximation with idempotent scalings are applicable to a large class of wavelet bi-frames.

In Chapter 7, we consider variational problems for noise removal from images. By applying the characterization of Besov spaces in Chapter 5, we derive an equivalent discrete variational problem in terms of wavelet bi-frame coefficients, and an approximate solution can be derived by thresholding the bi-frame expansion. In order to determine the threshold parameter, we apply the H-curve criterion to wavelet bi-frames. Recall that, in [MP03], it is only applied to an orthonormal wavelet bases. Finally, this chapter verifies that the method yields good results for a wavelet bi-frame as well. In comparison to the threshold choice according to the mean square error minimization, it turns out that the H-curve criterion provides better denoised images with respect to the visual perception. Hence, the numerical results are promising, and we finally solve (P4).

Chapter 1

The Classical Setting: Wavelet Bases

The present chapter is dedicated to a brief overview of the theory of wavelet bases in $L_2(\mathbb{R}^d)$, whose scope is versatile. On the one hand, the wavelet transform provides fast numerical algorithms which are successfully applied in signal and image processing. On the other hand, the approximation power of wavelets as well as their ability to characterize certain function spaces made them a valuable tool in pure and applied mathematics.

The success of wavelets is promoted by their inner properties, such as a high smoothness, a high number of vanishing moments, and small supports. However, these qualities are competing, and in order to construct wavelets, they require a careful balancing. In many applications, one also needs further inner properties. For instance, symmetric wavelets provide better results in image and signal analysis, cf. [Mal99].

Univariate orthonormal wavelets with compact support have been successfully constructed by Daubechies in her celebrated paper [Dau88], see also [Dau92]. Orthonormal wavelet bases seem most desirable since they provide Parseval's Equality. However, orthogonality is also very restrictive since it makes it hard or even impossible to find wavelets satisfying a variety of extra conditions such as symmetry.

In order to overcome these restrictions, one constructs two wavelet bases, a primal and a dual one, which are biorthogonal to each other. The weaker biorthogonal concept allows for symmetries, while still providing expansions similar to those of orthonormal wavelets. However, strong inner properties of primal wavelets generally lead to weak inner properties of dual wavelets. This limitation provides the motivation to the present work.

We proceed as follows: first, we introduce the concept of pairs of biorthogonal wavelet bases. Then, we address their construction based on a multiresolution analysis that is generated by a so-called refinable function. Within this framework, wavelets can be derived by finding bases for certain complementary spaces, and this search can actually be reduced to a matrix completion problem. Since the problem is often explicitly solvable, the construction of wavelet bases is reduced to the construction of refinable functions. Finally, we address the approximation order of biorthogonal wavelets, the fast wavelet transform, and the characterization of function spaces. Within this context, we discuss desirable inner properties of wavelets in detail. We conclude this chapter by pointing out restrictions and inflexibilities of orthogonal and biorthogonal wavelets.

1.1 Biorthogonal Wavelet Bases

1.1.1 Riesz Bases

Let $\mathcal{H}$ be a Hilbert space and $\mathcal{K}$ be some countable index set throughout. Then a collection $\{f_\kappa : \kappa \in \mathcal{K}\}$ is called *complete* in $\mathcal{H}$ if its linear span is dense. The collection $\{f_\kappa : \kappa \in \mathcal{K}\}$ is called an *orthonormal basis* if it is complete in $\mathcal{H}$ and satisfies the orthogonality relation

$$\langle f_\kappa, f_{\kappa'} \rangle = \delta_{\kappa,\kappa'}, \quad \text{for all } \kappa, \kappa' \in \mathcal{K}. \tag{1.1}$$

Then, each $f \in \mathcal{H}$ can be expanded by

$$f = \sum_{\kappa \in \mathcal{K}} \langle f, f_\kappa \rangle f_\kappa, \tag{1.2}$$

where the right-hand side converges *unconditionally*, i.e., the convergence does not depend on the ordering of $\mathcal{K}$. Due to Parseval's Theorem, the collection $\{f_\kappa : \kappa \in \mathcal{K}\}$ is an orthonormal basis iff its associated *synthesis operator*

$$F : \ell_2(\mathcal{K}) \to \mathcal{H}, \quad (c_\kappa)_{\kappa \in \mathcal{K}} \mapsto \sum_{\kappa \in \mathcal{K}} c_\kappa f_\kappa, \tag{1.3}$$

is unitary.

The orthonormality relations (1.1) are extremely strong inner properties. For instance, within the context of wavelets, they prohibit some desirable extra conditions, see Section 1.3 for details, and one can overcome some of those restrictions with a different concept:

Definition 1.1.1. A collection $\{f_\kappa : \kappa \in \mathcal{K}\}$ is called a *Riesz basis* for $\mathcal{H}$ if it is complete in $\mathcal{H}$ and there exist positive constants A, B such that, for all $(c_\kappa)_{\kappa \in \mathcal{K}} \in \ell_2(\mathcal{K})$,

$$A \left\| (c_\kappa)_{\kappa \in \mathcal{K}} \right\|_{\ell_2}^2 \leq \left\| \sum_{\kappa \in \mathcal{K}} c_\kappa f_\kappa \right\|_{\mathcal{H}}^2 \leq B \left\| (c_\kappa)_{\kappa \in \mathcal{K}} \right\|_{\ell_2}^2. \tag{1.4}$$

The constants A and B are called the *lower* and *upper Riesz bounds*, respectively.

Again, it turns out that the convergence

$$\sum_{\kappa \in \mathcal{K}} c_\kappa f_\kappa$$

in (1.4) is unconditional, cf. Corollary 3.2.5 in [Chr03]. Chapter 5 with Theorem 6.5.1 in [Chr03] imply the characterization of Riesz bases in terms of the synthesis operator:

Theorem 1.1.2. *The set $\{f_\kappa : \kappa \in \mathcal{K}\}$ is a Riesz basis iff its synthesis operator $F : \ell_2(\mathcal{K}) \to \mathcal{H}$ given by* (1.3) *is well-defined and invertible.*

In the case of Theorem 1.1.2, the Banach-Steinhaus Theorem implies that F is even bounded, see Lemma 3.2.1 in [Chr03]. Then the Open Mapping Theorem yields that F is boundedly invertible.

Due to Theorem 1.1.2, precisely the concept of Riesz bases provides a bijective correspondence between ℓ_2 and $\mathcal{H}$ such that each element f in $\mathcal{H}$ has a series expansion with coefficients in ℓ_2, i.e.,

$$f = \sum_{\kappa \in \mathcal{K}} c_\kappa f_\kappa, \quad (c_\kappa)_{\kappa \in \mathcal{K}} \in \ell_2(\mathcal{K}).$$

Contrary to an orthonormal basis, given $f \in \mathcal{H}$, the coefficients are, in general, not the inner products $(\langle f, f_\kappa \rangle)_{\kappa \in \mathcal{K}}$. Nevertheless, for each Riesz basis $\{f_\kappa : \kappa \in \mathcal{K}\}$, there exists a second Riesz basis $\{\widetilde{f}_\kappa : \kappa \in \mathcal{K}\}$, which is biorthogonal to $\{f_\kappa : \kappa \in \mathcal{K}\}$, i.e.,

$$\left\langle f_\kappa, \widetilde{f}_{\kappa'} \right\rangle = \delta_{\kappa,\kappa'}, \quad \text{for all } \kappa, \kappa' \in \mathcal{K},$$

see Theorem 3.6.3 in [Chr03]. Then each $f \in \mathcal{H}$ has the series expansion

$$f = \sum_{\kappa \in \mathcal{K}} \left\langle f, \widetilde{f}_\kappa \right\rangle f_\kappa, \tag{1.5}$$

and we say that $\{f_\kappa : \kappa \in \mathcal{K}\}$, $\{\widetilde{f}_\kappa : \kappa \in \mathcal{K}\}$ constitute a pair of *biorthogonal Riesz bases*. Hence, the biorthogonal concept is much weaker than the orthogonal one, but it still provides by (1.5) an expansion similar to (1.2).

1.1.2 Wavelets with General Dilation Matrices

First, we shall clarify our concept of dilation. Throughout this thesis, let M denote a *dilation matrix*, i.e., an integer matrix, whose eigenvalues are greater than one in modulus. In case $M = 2\mathcal{I}_d$, we speak of *dyadic* dilation, and the canonical univariate choice is $M = 2$. Moreover, let us have a closer look at two subclasses of dilation matrices. A dilation matrix is called *isotropic* if it can be diagonalized and all eigenvalues have the same modulus. This class is mainly addressed in Chapter 5 of the present thesis. A dilation matrix is called *idempotent* if there are $l, h \in \mathbb{N}$ such that

$$M^l = h\mathcal{I}_d.$$

Idempotent dilation matrices are of main interest in Section 6.2.2. It turns out that the second class is contained in the first one, see Appendix A.2 for the proof of the following lemma:

Lemma 1.1.3. *Each idempotent dilation matrix is isotropic.*

In the bivariate setting, the two popular dilation matrices

$$M_b = \begin{pmatrix} 1 & 1 \\ 1 & -1 \end{pmatrix}, \quad M_q = \begin{pmatrix} 1 & -1 \\ 1 & 1 \end{pmatrix} \tag{1.6}$$

are called *box spline matrix* and *quincunx matrix*, respectively. Since they satisfy $M_b^2 = 2\mathcal{I}_2$ and $M_q^4 = -4\mathcal{I}_2$, they are idempotent. Moreover, both matrices generate the *quincunx grid*, i.e.,

$$M_b\mathbb{Z}^2 = M_q\mathbb{Z}^2 = \left\{k \in \mathbb{Z}^2 : k_1 + k_2 \in 2\mathbb{Z}\right\},$$

see Figure 1.1.

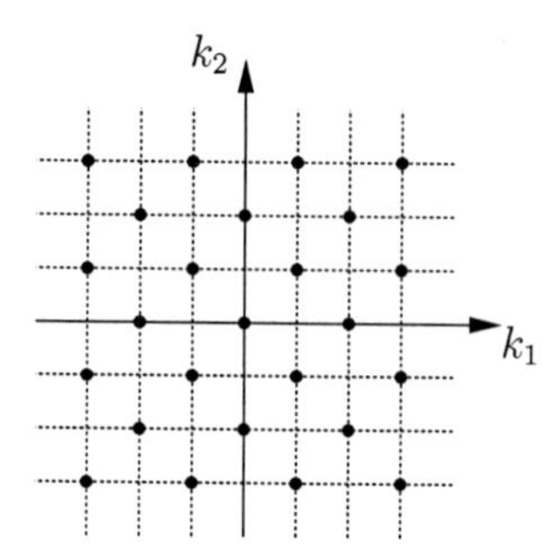

Figure 1.1: The quincunx grid

So far, we introduced our concept of dilation. In wavelet analysis, one considers dilates and shifts of functions, i.e., for $f : \mathbb{R}^d \to \mathbb{C}$, we address

$$f_{j,k}(x) := m^{\frac{j}{2}} f(M^j x - k), \quad \text{for } j \in \mathbb{Z}, k \in \mathbb{Z}^d,$$

where $m := |\det(M)|$ throughout. Then given a finite number of $L_2(\mathbb{R}^d)$-functions $\psi^{(1)}, \dots, \psi^{(n)}$, the collection

$$X(\{\psi^{(1)}, \dots, \psi^{(n)}\}) := \left\{\psi_{j,k}^{(\mu)} : j \in \mathbb{Z}, k \in \mathbb{Z}^d, \mu = 1, \dots, n\right\} \tag{1.7}$$

is called a *wavelet system*, and the functions $\psi^{(1)}, \dots, \psi^{(n)}$ are called *wavelets*. However, Gröchenig writes in [Grö01],

> "The terminology is a bit confusing because there is no general accepted definition of a wavelet. ..., but almost any function has been called a wavelet at some time or other."

If we speak of a basis in the wavelet context, then we mean a Riesz basis throughout:

Definition 1.1.4. Two wavelet systems $X(\{\psi^{(1)}, \dots, \psi^{(n)}\})$, $X(\{\widetilde{\psi}^{(1)}, \dots, \widetilde{\psi}^{(n)}\})$ are called a *pair of biorthogonal wavelet bases* if they constitute a pair of biorthogonal Riesz bases in $L_2(\mathbb{R}^d)$.

Compatible to Definition 1.1.4, a wavelet system $X(\{\psi^{(1)}, \dots, \psi^{(n)}\})$ is called an *orthonormal wavelet basis* if it constitutes an orthonormal basis in $L_2(\mathbb{R}^d)$. Next, we present some examples. They are verified in Section 1.1.3.

Example 1.1.5. In the univariate dyadic setting, the *Haar wavelet*

$$\psi_H = 1_{[\frac{1}{2},1)} - 1_{[0,\frac{1}{2})} \tag{1.8}$$

yields an orthonormal basis $X(\{\psi_H\})$ for $L_2(\mathbb{R})$. In fact, this was known long before the development of wavelet theory.

The following example provides an orthonormal wavelet basis with the box spline matrix.

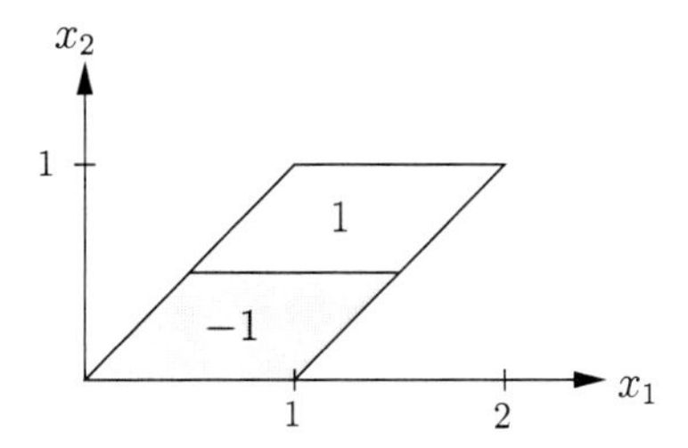

Figure 1.2: The wavelet ψ in Example 1.1.6. It is equal to 1 on the upper parallelogram, equal to -1 on the lower one, and 0 elsewhere.

Example 1.1.6. Let $M = M_b$ be the box spline dilation matrix. Given ψ as in Figure 1.2, the wavelet system $X(\{\psi\})$ constitutes an orthonormal basis for $L_2(\mathbb{R}^2)$.

Example 1.1.6 can be obtained from the theory of self-similar tilings as we shall explain in the sequel. Given a dilation matrix M and Γ_M^* a complete set of representatives of $\mathbb{Z}^d/M\mathbb{Z}^d$, let

$$Q := \Big\{ \sum_{n=1}^{\infty} M^{-n}\gamma_n^* : \gamma_n^* \in \Gamma_M^* \Big\}. \tag{1.9}$$

The Lebesgue measure of Q is an integer, and Q is *self-affine* with respect to M and Γ_M^*, i.e.,

$$Q = \bigcup_{\gamma^* \in \Gamma_M^*} \left(M^{-1}Q + \gamma^*\right),$$

where the union is disjoint up to a set of measure zero, see [GM92] for details. If the Lebesgue measure of Q is equal to one, then the characteristic function 1_Q has orthonormal integer shifts in $L_2(\mathbb{R}^d)$, i.e.,

$$\langle 1_Q(\cdot - k), 1_Q(\cdot - l)\rangle = \delta_{k,l}, \quad \text{for } k, l \in \mathbb{Z}^d,$$

cf. [GM92]. In case of Example 1.1.6, let $\varphi = 1_Q$ be the characteristic function of the self-affine set Q in (1.9) with respect to M and $\Gamma_M^* = \{0, (1,0)^\top\}$. It turns out that Q equals the union of both grey parallelograms in Figure 1.2. Then one easily verifies

$$\psi(x) = \varphi(M_b x - (1,0)^\top) - \varphi(M_b x). \tag{1.10}$$

A similar relation holds in Example 1.1.5: the univariate dyadic situation allows for $\Gamma_M^* = \{0, 1\}$, then Q given by (1.9) is equal to $[0,1)$, and the Haar wavelet ψ_H satisfies

$$\psi_H(x) = 1_{[0,1)}(2x - 1) - 1_{[0,1)}(2x). \tag{1.11}$$

The following example starts with the function φ:

Example 1.1.7. Let $M = M_q$ be the quincunx dilation matrix. For $\Gamma_M^* = \{0, (1,0)^\top\}$, the self-affine set Q in (1.9) is given in Figure 1.3, and it is called the *twin-dragon*. Let $\varphi = 1_Q$, and let

$$\psi(x) := \varphi(M_q x - (1,0)^\top) - \varphi(M_q x). \tag{1.12}$$

Then $X(\{\psi\})$ constitutes an orthonormal basis for $L_2(\mathbb{R}^2)$.

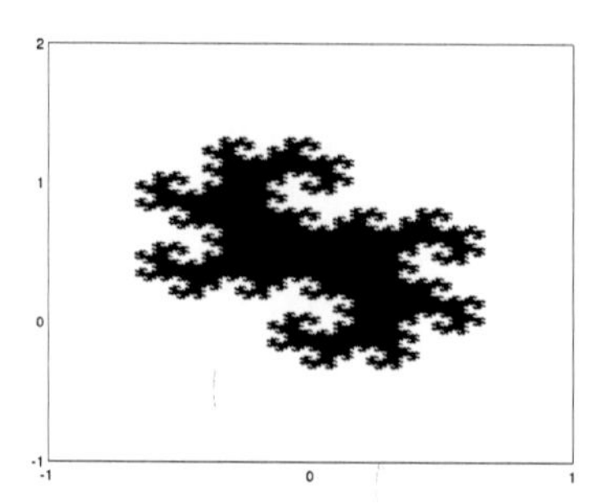

Figure 1.3: The twin-dragon in Example 1.1.7. It is fractal, but has the Lebesgue measure of one.

Remark 1.1.8. The identities (1.10), (1.11), and (1.12) correspond to a general construction principle, which we present in the following section. There, we also verify Examples 1.1.5, 1.1.6, and 1.1.7.

The self-affine set generated by the box spline matrix M_b is a parallelogram. Hence, it is somehow more regular than the fractal twin-dragon in Example 1.1.7. This observation provides a clue that the two matrices behave quite differently in the context of wavelets. The box spline matrix M_b allows for arbitrarily smooth compactly supported orthonormal wavelet bases. For compactly supported smooth wavelets with the quincunx matrix M_q, as far as we know, we have to switch into the weaker concept of biorthogonal wavelets, cf. Section 1.3.

1.1.3 The Multiresolution Analysis

Mallat and Meyer proposed in [Mal89, Mey90] the concept of multiresolution analysis, which is a powerful framework for the construction of wavelets. Since its first dyadic appearance, several generalizations have been developed. We recall the concept with respect to general dilation matrices in $\mathbb{R}^d$:

Definition 1.1.9. An increasing sequence of closed subspaces $(V_j)_{j\in\mathbb{Z}}$ in $L_2(\mathbb{R}^d)$ is called a *multiresolution analysis* if the following holds:

(M-1) $f \in V_j$ iff $f(M^{-j}\cdot) \in V_0$, for all $j \in \mathbb{Z}$,

(M-2) $\bigcup_{j\in\mathbb{Z}} V_j$ is dense in $L_2(\mathbb{R}^d)$,

(M-3) $\bigcap_{j\in\mathbb{Z}} V_j = \{0\}$,

(M-4) there is a function $\varphi \in V_0$, whose integer shifts constitute a Riesz basis for V_0.

The function φ in (M-4) is called the *generator* of the multiresolution analysis.

It should be mentioned that a function in $L_2(\mathbb{R}^d)$ is called *stable* if its integer shifts constitute a Riesz basis for their closed linear span. Hence, (M-4) requires that φ is stable.

In order to construct compactly supported biorthogonal wavelets, let $(V_j)_{j\in\mathbb{Z}}$ and $(\widetilde{V}_j)_{j\in\mathbb{Z}}$ be two multiresolution analyses with compactly supported generators φ and $\widetilde{\varphi}$, respectively. In addition, we suppose that their integer shifts are biorthogonal to each other, i.e., for all $k, k' \in \mathbb{Z}^d$,

$$\langle \varphi(\cdot - k), \widetilde{\varphi}(\cdot - k') \rangle = \delta_{k,k'}. \tag{1.13}$$

Let W_0 and $\widetilde{W}_0$ be complementary spaces of V_0 in V_1 and of $\widetilde{V}_0$ in $\widetilde{V}_1$, i.e,

$$V_1 = V_0 \oplus W_0, \quad \widetilde{V}_1 = \widetilde{V}_0 \oplus \widetilde{W}_0. \tag{1.14}$$

Moreover, they are taken to be related by

$$W_0 \perp \widetilde{V}_0, \quad \widetilde{W}_0 \perp V_0. \tag{1.15}$$

Then, one has to find wavelets $\psi^{(\mu)}$ and $\widetilde{\psi}^{(\mu)}$, $\mu = 1, \ldots, n$, such that their integer shifts constitute Riesz bases of W_0 and $\widetilde{W}_0$, respectively. According to the theory of shift invariant spaces, the number of wavelets is determined by $n = m - 1$, see also the textbook [Woj97]. In order to derive birothogonal wavelets, they have to be choosen such that

$$\left\langle \psi_{0,k}^{(\mu)}, \widetilde{\psi}_{0,k'}^{(\mu')} \right\rangle = \delta_{k,k'}\delta_{\mu,\mu'}. \tag{1.16}$$

So far, we have biorthogonality on the scale $j = 0$. According to the multiresolution analysis framework, this geometrical relation extends without any more effort, as we shall explain in the following. Once we have found W_0 and $\widetilde{W}_0$, the definitions

$$\begin{aligned} f \in W_j \quad &\text{iff} \quad f(M^{-j}\cdot) \in W_0, \\ f \in \widetilde{W}_j \quad &\text{iff} \quad f(M^{-j}\cdot) \in \widetilde{W}_0 \end{aligned}$$

provide two sequences of subspaces $(W_j)_{j\in\mathbb{Z}}$ and $(\widetilde{W}_j)_{j\in\mathbb{Z}}$. They share the multiresolution analysis structure (M-1), and they extend (1.14) and (1.15) to each scale $j \in \mathbb{Z}$, i.e.,

$$V_{j+1} = V_j \oplus W_j, \quad \widetilde{V}_{j+1} = \widetilde{V}_j \oplus \widetilde{W}_j \tag{1.17}$$

and

$$W_j \perp \widetilde{V}_j, \quad \widetilde{W}_j \perp V_j. \tag{1.18}$$

For fixed $j \in \mathbb{Z}$, the collections

$$\left\{\psi_{j,k}^{(\mu)} : \mu = 1, \ldots, m-1,\ k \in \mathbb{Z}^d\right\}, \quad \left\{\widetilde{\psi}_{j,k}^{(\mu)} : \mu = 1, \ldots, m-1,\ k \in \mathbb{Z}^d\right\}$$

are Riesz bases for W_j and $\widetilde{W}_j$, respectively, and we finally obtain the complete biorthogonality relations

$$\left\langle \psi_{j,k}^{(\mu)}, \widetilde{\psi}_{j',k'}^{(\mu')} \right\rangle = \delta_{j,j'}\delta_{k,k'}\delta_{\mu,\mu'}. \tag{1.19}$$

According to (M-3), the relations in (1.17) yield the decompositions up to level j,

$$V_j = \bigoplus_{j'=-\infty}^{j-1} W_{j'}, \quad \widetilde{V}_j = \bigoplus_{j'=-\infty}^{j-1} \widetilde{W}_{j'},$$

and by applying (M-2) and (1.17), one derives also the complete decompositions

$$L_2(\mathbb{R}^d) = \bigoplus_{j\in\mathbb{Z}} W_j, \quad L_2(\mathbb{R}^d) = \bigoplus_{j\in\mathbb{Z}} \widetilde{W}_j, \tag{1.20}$$

see [Dau92] and [CT97] for details. Thus, the wavelet systems

$$X(\{\psi^{(1)},\dots,\psi^{(m-1)}\}), \quad X(\{\widetilde{\psi}^{(1)},\dots,\widetilde{\psi}^{(m-1)}\}) \tag{1.21}$$

are biorthogonal to each other, and they are complete in $L_2(\mathbb{R}^d)$.

In order to turn the above framework into a more applicable form, we have a closer look at a multiresolution analysis. Given some generator φ, according to (M-1) and (M-4), the collection

$$\left\{\varphi_{j,k} : k \in \mathbb{Z}^d\right\}$$

is a Riesz basis for V_j. Since the spaces V_j are increasing, φ is contained in V_1. Thus, there exists a sequence $(a_k)_{k\in\mathbb{Z}^d} \in \ell_2(\mathbb{Z}^d)$ such that φ satisfies the *refinement equation*

$$\varphi(x) = \sum_{k\in\mathbb{Z}} a_k \varphi(Mx - k). \tag{1.22}$$

Therefore, we call φ *refinable*, and the sequence $(a_k)_{k\in\mathbb{Z}^d}$ is called its *mask* or its *filter*. Since we focus on compactly supported wavelets, it is reasonable that we suppose that the generator φ has compact support and that its mask is finitely supported. It should be mentioned that among the collection of compactly supported distributions, the solution of the refinement equation is unique up to multiplication with a constant, cf. [CDM91]. Applying the Fourier transform to both sides of (1.22) yields

$$\widehat{\varphi}(\xi) = a(M^{-\top}\xi)\widehat{\varphi}(M^{-\top}\xi), \tag{1.23}$$

where the trigonometric polynomial

$$a(\xi) = \frac{1}{m} \sum_{k\in\mathbb{Z}^d} a_k e^{-2\pi i k\cdot\xi}$$

is called the *symbol* of φ, see Appendix A.1 for the normalization of the Fourier transform.

Remark 1.1.10. Throughout this thesis, symbols are trigonometric polynomials, and hence, their coefficients are finitely supported sequences. It should be mentioned that the term symbol sometimes includes arbitrary $\mathbb{Z}^d$-periodic functions in the literature.

The iteration of (1.23) yields

$$\widehat{\varphi}(\xi) = \prod_{j=1}^{l} a(M^{\top^{-j}}\xi)\widehat{\varphi}(M^{\top^{-l}}\xi), \quad \text{for } l \in \mathbb{N}. \tag{1.24}$$

Since $M^{\top^{-j}}\xi$ tends to zero as j goes to infinity, we are tempted to consider the limit. The Fourier transform of φ is continuous in zero, because φ is compactly supported. Hence, the convergence in (1.24) for $l \to \infty$ requires $a(0) = 1$. This is also sufficient as we shall see in the following, where we have a different starting point, and we reverse the process described above.

Given some symbol a with $a(0) = 1$, it induces a multiresolution analysis in the following way. We define φ by its Fourier transform

$$\widehat{\varphi}(\xi) = \prod_{j\geq 1} a\big(M^{\top^{-j}}\xi\big). \tag{1.25}$$

According to [Dau92], the right-hand side converges uniformly on compact sets, and φ is a compactly supported distribution, normalized by $\widehat{\varphi}(0) = 1$, see Appendix A.1 for distributions. It satisfies the refinement equation (1.22), at least in the distributional sense. If φ is contained in $L_2(\mathbb{R}^d)$ and stable, then we can define V_0 by (M-4). A sequence of closed subspaces $(V_j)_{j\in\mathbb{Z}}$ is derived by applying (M-1) as a definition for V_j. Since φ is refinable, the subspaces are increasing and, according to [dBDR93], they constitute a multiresolution analysis. Hence, we have obtained a refinable function and an underlying multiresolution analysis by a suitable choice of some symbol.

Let a and b be two symbols with $a(0) = b(0) = 1$ generating refinable functions φ and $\widetilde{\varphi}$, respectively. We suppose that both are contained in $L_2(\mathbb{R}^d)$ and that they are stable. Let us denote the generated multiresolution analyses by $(V_j)_{j\in\mathbb{Z}}$ and $(\widetilde{V}_j)_{j\in\mathbb{Z}}$, respectively. Then the biorthogonality relation (1.13) is equivalent to

$$\sum_{\gamma\in\Gamma_M} a(\xi+\gamma)\overline{b(\xi+\gamma)} = 1, \quad \text{for all } \xi \in \mathbb{R}^d, \tag{1.26}$$

where Γ_M is a complete set of representatives of $M^{-\top}\mathbb{Z}^d/\mathbb{Z}^d$ with $0 \in \Gamma_M$ throughout this work, cf. [Dau92]. If (1.26) holds, then b is called a *dual symbol* of a. Since $\{\varphi_{1,k} : k \in \mathbb{Z}^d\}$ and $\{\widetilde{\varphi}_{1,k} : k \in \mathbb{Z}^d\}$ are Riesz bases of V_1 and $\widetilde{V}_1$, respectively, the inclusions $W_0 \subset V_1$ and $\widetilde{W}_0 \subset \widetilde{V}_1$ provide that there exist sequences $\big(a_k^{(\mu)}\big)_{k\in\mathbb{Z}^d}$ and $\big(b_k^{(\mu)}\big)_{k\in\mathbb{Z}^d}$ such that

$$\psi^{(\mu)}(x) = \sum_{k\in\mathbb{Z}^d} a_k^{(\mu)}\varphi(Mx-k), \tag{1.27}$$

$$\widetilde{\psi}^{(\mu)}(x) = \sum_{k\in\mathbb{Z}^d} b_k^{(\mu)}\widetilde{\varphi}(Mx-k). \tag{1.28}$$

The coefficient sequences are necessarily contained in $\ell_2(\mathbb{Z}^d)$. In order to derive compactly supported wavelets, we try to choose finitely supported sequences. Provided that we are successful, applying the Fourier transform to (1.27) and (1.28) yields

$$\widehat{\psi^{(\mu)}}(\xi) = a^{(\mu)}(M^{-\top}\xi)\widehat{\varphi}(M^{-\top}\xi), \tag{1.29}$$

$$\widehat{\widetilde{\psi}^{(\mu)}}(\xi) = b^{(\mu)}(M^{-\top}\xi)\widehat{\widetilde{\varphi}}(M^{-\top}\xi), \tag{1.30}$$

where $a^{(\mu)}$ and $b^{(\mu)}$ denote the symbols according to the finitely supported sequences $\big(a_k^{(\mu)}\big)_{k\in\mathbb{Z}^d}$ and $\big(b_k^{(\mu)}\big)_{k\in\mathbb{Z}^d}$, respectively. The geometrical conditions (1.13), (1.15), and (1.16) with the complement property (1.14) imply

$$\sum_{\gamma\in\Gamma_M} a^{(\mu)}(\xi+\gamma)\overline{b^{(\nu)}(\xi+\gamma)} = \delta_{\mu,\nu}, \quad \mu,\nu = 0,\dots,m-1, \tag{1.31}$$

where $a^{(0)} := a$ and $b^{(0)} := b$, see [Dau92] for details. Note that (1.31) includes the duality relations (1.26).

The following theorem turns the ideas above into a construction concept for pairs of compactly supported biorthogonal wavelet bases. It tells us that the necessary conditions (1.31) are already sufficient. In [RS97b], the theorem is obtained under a mild smoothness assumption on the generators that can be removed by the results in [Bow00, CSS98].

Theorem 1.1.11. *Given a symbol a and a dual symbol b with $a(0) = b(0) = 1$, let them generate two stable refinable functions φ, $\widetilde{\varphi} \in L_2(\mathbb{R}^d)$, respectively. Given additional symbols $a^{(\mu)}$ and $b^{(\mu)}$, $\mu = 1, \dots, m-1$, satisfying the conditions* (1.31) *as well as*

$$a^{(\mu)}(0) = b^{(\mu)}(0) = 0, \quad \textit{for all } \mu = 1, \dots, m-1, \tag{1.32}$$

we define $\psi^{(\mu)}$ and $\widetilde{\psi}^{(\mu)}$, $\mu = 1, \dots, m-1$, by (1.27) *and* (1.28), *respectively. Then the systems*

$$X(\{\psi^{(1)}, \dots, \psi^{(m-1)}\}), \quad X(\{\widetilde{\psi}^{(1)}, \dots, \widetilde{\psi}^{(m-1)}\})$$

constitute a pair of compactly supported biorthogonal wavelet bases.

Theorem 1.1.11 is some good news for the construction of biorthogonal wavelets. First, one chooses a symbol a and a dual symbol b. Then one needs to verify membership in $L_2(\mathbb{R}^d)$ and stability of the generated refinable functions. Advantageously, these properties can be ensured by certain conditions on the symbols, cf. [Dau92]. Once these ingredients are established, the construction of wavelets simply requires the choice of wavelet symbols satisfying some zero condition and (1.31).

By applying Theorem 1.1.11, we can verify the three Examples 1.1.5, 1.1.6, and 1.1.7, which provide orthonormal wavelet bases. Recall that a symbol is called *orthogonal* if it is dual to itself, i.e.,

$$\sum_{\gamma \in \Gamma_M} |a(\xi + \gamma)|^2 = 1, \quad \text{for all } \xi \in \mathbb{R}^d.$$

First, we derive the Haar wavelet from the multiresolution analysis approach:

Example 1.1.12. In the univariate dyadic setting, let

$$a(\xi) = \frac{1 + e^{-2\pi i \xi}}{2}.$$

Then a is orthogonal, and it generates $\varphi = 1_{[0,1)}$. For

$$a^{(1)}(\xi) := e^{-2\pi i \xi} \overline{a(\xi + \tfrac{1}{2})},$$

the conditions (1.31) hold with $a = b$ and $a^{(1)} = b^{(1)}$. Then

$$\psi(x) = \sum_{k \in \mathbb{Z}} a_k^{(1)} 1_{[0,1)}(2x - k)$$

is exactly the same as (1.11), and ψ is the Haar wavelet (1.8). According to Theorem 1.1.11, the system $X(\{\psi\})$ constitutes an orthonormal basis for $L_2(\mathbb{R})$.

Next, we verify Example 1.1.6:

Example 1.1.13. Given the box spline dilation matrix $M = M_b$, let the bivariate symbol a be defined by

$$a(\xi) = \frac{1 + e^{-2\pi i \xi_1}}{2}. \tag{1.33}$$

Then a is orthogonal, and it generates the characteristic function φ of the union of both parallelograms in Figure 1.2, cf. [GM92]. For

$$a^{(1)}(\xi) := e^{-2\pi i \xi_1} \overline{a\big(\xi + (\tfrac{1}{2}, \tfrac{1}{2})^\top\big)}, \tag{1.34}$$

the conditions (1.31) hold with $a = b$ and $a^{(1)} = b^{(1)}$. Then

$$\psi(x) = \sum_{k \in \mathbb{Z}^2} a_k^{(1)} \varphi(M_b x - k)$$

is nothing other than (1.10), and ψ equals the wavelet in Example 1.1.6. According to Theorem 1.1.11, $X(\{\psi\})$ constitutes an orthonormal wavelet basis.

Given a collection of symbols, different choices of the dilation matrix can yield different refinable functions and wavelets. Replacing M_b by M_q in Example 1.1.13 yields the twin-dragon of Example 1.1.7:

Example 1.1.14. Let $M = M_q$ be the quincunx dilation matrix. By applying $\Gamma_{M_b} = \Gamma_{M_q}$, the symbol a in (1.33) is also orthogonal with respect to the dilation matrix M_q. It generates the characteristic function φ of the twin-dragon in Figure 1.3, cf. [GM92]. By the choice of $a^{(1)}$ as in (1.34), the conditions (1.31) hold and

$$\psi(x) = \sum_{k \in \mathbb{Z}^2} a_k^{(1)} \varphi(M_q x - k)$$

is nothing other than (1.12). Then ψ equals the wavelet in Example 1.1.7. According to Theorem 1.1.11, $X(\{\psi\})$ constitutes an orthonormal wavelet basis.

Another way to obtain multivariate wavelets is applying tensor products to univariate systems, see [Dau92] for the following example:

Example 1.1.15. Given a univariate dyadic orthogonal wavelet basis $X(\{\psi\})$ with compact support and underlying refinable function φ, then the system

$$X(\{\varphi \otimes \psi, \psi \otimes \varphi, \psi \otimes \psi\}) \tag{1.35}$$

is a bivariate dyadic orthogonal wavelet basis with underlying refinable function $\varphi \otimes \varphi$.

Wavelet systems like (1.35) are called *separable* because each wavelet is a tensor product of univariate functions. Due to the tensor structure, separable wavelets "prefer" the axis directions. Especially in image processing, this is quite inconvenient. Then nonseparable bases avoid such directional preferences, and they provide better results, cf. [Mal89].

Similar to Example 1.1.15, multivariate bases can be obtained by multiple tensor products of univariate orthonormal bases. However, this provides similar directional dependencies as the bivariate tensor approach. Moreover, the number of wavelets is $2^d - 1$, which grows exponentially, and this causes complexity problems in applications. In the context of the fast wavelet transform, we address this topic in Subsection 1.2.2.

In Examples 1.1.13 and 1.1.14, we already derived bivariate nonseparable wavelets. However, they are not even continuous. In the sequel, we discuss some smooth, multivariate, nonseparable refinable functions. Following [dBHR93], we introduce box splines in arbitrary dimensions that may lead to biorthogonal wavelet bases. For a fixed integer $n \geq d$, given *direction vectors*

$$y^{(1)}, \ldots, y^{(n)} \in \mathbb{Z}^d,$$

let $Y_{n'}$, $d \leq n' \leq n$ be the matrix of the first n' vectors, i.e,

$$Y_{n'} = \left(y^{(1)}, \ldots, y^{(n')}\right),$$

while we suppose $\det(Y_d) \neq 0$. Then the *box spline* φ_{Y_n} with respect to the *direction matrix* Y_n is recursively defined by

$$\varphi_{Y_{n'}}(x) = \int_0^1 \varphi_{Y_{n'-1}}(x - ty^{(n')})dt, \quad \text{for all } d < n' \leq n,$$
$$\varphi_{Y_d} = \frac{1}{|\det(Y_d)|} 1_{Y_d[0,1)^d}.$$

The box spline is refinable with respect to dyadic dilation, and its smoothness can easily be read off the direction matrix, cf. [dBHR93]:

Lemma 1.1.16. *Let Y_n be some direction matrix. Then the following holds:*

(a) *The box spline φ_{Y_n} is refinable with respect to the symbol*

$$a_{Y_n}(\xi) = \prod_{\nu=1}^{n} \frac{1 + e^{-2\pi i y^{(\nu)} \cdot \xi}}{2}$$

and dyadic dilation.

(b) *Given an integer $\alpha \geq 2$ such that there exist $n - \alpha + 1$ linearly independent column vectors in Y_n, then φ_{Y_n} is $\alpha - 2$ times differentiable.*

In order to ensure that φ_{Y_n} is a generator of a multiresolution analysis, we still have to verify stability. Advantageously, it can also be read off the direction matrix, see [dBHR93].

Lemma 1.1.17. *Given a box spline φ_{Y_n}, the following statements are equivalent:*

(i) *φ_{Y_n} is stable,*

(ii) *Y_n is unimodular, i.e.,*

$$|\det(Y)| \in \{0, 1\},$$

for all $d \times d$-submatrices Y of Y_n.

Thus, given a unimodular direction matrix, the box spline φ_{Y_n} generates a multiresolution analysis. Since the construction of a pair of biorthogonal wavelet bases requires an additional second generator, we also need a dual symbol of a_{Y_n}. Its existence can be ensured by a general concept of linear independence: we say a compactly supported distribution φ has *globally linearly independent integer shifts* if the mapping

$$F : \ell(\mathbb{Z}^d) \to \mathcal{S}'(\mathbb{R}^d), \quad (\lambda_k)_{k\in\mathbb{Z}^d} \mapsto \sum_{k\in\mathbb{Z}^d} \lambda_k \varphi(\cdot - k) \tag{1.36}$$

is injective, where $\mathcal{S}'(\mathbb{R}^d)$ denotes the space of tempered distributions, cf. Appendix A.1. According to a result in [DM97], if φ has globally linearly independent integer shifts, then its symbol a has a dual symbol b.

In the box spline setting, global linear independence and stability are equivalent, see [dBHR93] for the following extension of Lemma 1.1.17.

Lemma 1.1.18. *Given a box spline φ_{Y_n}, the following statements are equivalent:*

(i) *φ_{Y_n} has globally linearly independent integer shifts,*

(ii) *φ_{Y_n} is stable,*

(iii) *Y_n is unimodular, i.e.,*

$$|\det(Y)| \in \{0, 1\},$$

for all $d \times d$-submatrices Y of Y_n.

Thus, given a_{Y_n} with unimodular direction matrix Y_n, then φ_{Y_n} generates a multiresolution analysis, and there exists a dual symbol b of a_{Y_n}. For the application of Theorem 1.1.11, we still need additional symbols such that (1.31) holds. Since this problem does not only arise in box spline constructions, we address the topic of finding additional symbols in more generality in the following subsection.

1.1.4 A Matrix Completion Problem

Let $\Gamma_M = \{0, \gamma_1, \ldots, \gamma_{m-1}\}$ denote a complete set of reprasentatives of $M^{-\top}\mathbb{Z}^d/\mathbb{Z}^d$. Then given symbols $a^{(\mu)}$ and $b^{(\mu)}$, $\mu = 0, \ldots, m-1$, the square matrices

$$\mathbf{a} := \left(a^{(\mu)}(\cdot + \gamma_\nu)\right)_{\substack{\nu=0,\ldots,m-1\\ \mu=0,\ldots,m-1}}, \qquad \mathbf{b} := \left(b^{(\mu)}(\cdot + \gamma_\nu)\right)_{\substack{\nu=0,\ldots,m-1\\ \mu=0,\ldots,m-1}} \tag{1.37}$$

are called *modulation matrices*. Due to the $\mathbb{Z}^d$-periodicity of trigonometric polynomials, (1.31) can be rewritten into

$$\mathbf{a}^\top \overline{\mathbf{b}} = \mathcal{I}_m. \tag{1.38}$$

Given only a symbol $a^{(0)}$ and a dual symbol $b^{(0)}$, the application of Theorem 1.1.11 requires additional symbols such that (1.38) holds. In other words, we must complete the matrices $\mathbf{a}$ and $\mathbf{b}$. The existence of a completion can be ensured by the so-called Quillen-Suslin Theorem, cf. [Qui76, Sus76], as we shall explain next. The theorem is not directly applicable since the columns of modulation matrices are highly redundant and each entry already determines its entire column. In the following, we transform modulation matrices into matrices with decoupled columns. This allows for the application of the theorem, and the completion of the decoupled system also provides a completion of the modulation matrices.

Given a symbol a, we denote its γ^*-*subsymbol* by

$$A_{\gamma^*}(\xi) := \sum_{k\in\mathbb{Z}^d} a_{Mk+\gamma^*} e^{-2\pi i k\cdot\xi}, \tag{1.39}$$

where

$$\Gamma_M^* = \{0, \gamma_1^*, \dots, \gamma_{m-1}^*\}$$

is a complete set of representatives of $\mathbb{Z}^d/M\mathbb{Z}^d$. Hence, a can be decomposed into

$$a(\xi) = \frac{1}{m} \sum_{\gamma^*\in\Gamma_M^*} A_{\gamma^*}(M^\top\xi) e^{-2\pi i\gamma^*\cdot\xi}. \tag{1.40}$$

An application of a result about character sums, i.e.,

$$\sum_{\gamma\in\Gamma_M} e^{2\pi i k\cdot\gamma} = \begin{cases} m, & \text{if } k \in M\mathbb{Z}^d, \\ 0, & \text{otherwise}, \end{cases} \tag{1.41}$$

provides the computation of the subsymbols from the Γ_M-shifts of a by

$$A_{\gamma^*}(M^\top\xi) = \sum_{\gamma\in\Gamma_M} e^{2\pi i\gamma^*\cdot(\xi+\gamma)} a(\xi+\gamma), \tag{1.42}$$

see [CL94] for details. For symbols $a^{(\mu)}$ and $b^{(\mu)}$, $\mu = 0, \dots, m-1$, we denote their subsymbols by $A_{\gamma_\nu^*}^{(\mu)}$ and $B_{\gamma_\nu^*}^{(\mu)}$, $\nu = 0, \dots, m-1$, respectively. Then the two matrices

$$\mathbf{A} = \left(A_{\gamma_\nu^*}^{(\mu)}\right)_{\substack{\nu=0,\dots,m-1\\ \mu=0,\dots,m-1}}, \qquad \mathbf{B} = \left(B_{\gamma_\nu^*}^{(\mu)}\right)_{\substack{\nu=0,\dots,m-1\\ \mu=0,\dots,m-1}}$$

are called *polyphase matrices*. Let

$$U(\xi) := \left(e^{-2\pi i\gamma_\mu^*\cdot(\xi+\gamma_\nu)}\right)_{\substack{\nu=0,\dots,m-1\\ \mu=0,\dots,m-1}}, \tag{1.43}$$

then by applying (1.40), we obtain

$$\mathbf{a}(\xi) = \frac{1}{m} U(\xi)\mathbf{A}(M^\top\xi), \tag{1.44}$$

$$\mathbf{b}(\xi) = \frac{1}{m} U(\xi)\mathbf{B}(M^\top\xi). \tag{1.45}$$

Moreover, (1.41) yields that $\frac{1}{\sqrt{m}}U$ is unitary. This implies

$$\mathbf{a}^\top\overline{\mathbf{b}} = \mathcal{I}_m \quad \text{iff} \quad \mathbf{A}^\top\overline{\mathbf{B}} = m\mathcal{I}_m. \tag{1.46}$$

At this point the famous theorem of Quillen-Suslin is applicable. In its full generality, it states that every projective module over some polynomial ring is free, see [Qui76, Sus76]. In [Swa78], Swan extended the result to Laurent polynomial rings. According to the identification of $\sum_{k\in\mathbb{Z}^d} a_k e^{-2\pi i k\cdot\xi}$ with $\sum_{k\in\mathbb{Z}^d} a_k z^k$, the result is also applicable to trigonometric polynomials. In order to avoid all of the algebraic background, we only explain the consequences in our setting. Given a symbol $a^{(0)}$ and a dual symbol $b^{(0)}$, then their subsymbols satisfy

$$\left(A^{(0)}_{\gamma_0^*}, \ldots, A^{(0)}_{\gamma_{m-1}^*}\right) \cdot \left(\overline{B^{(0)}_{\gamma_0^*}, \ldots, B^{(0)}_{\gamma_{m-1}^*}}\right)^\top = m. \tag{1.47}$$

In other words, the ideal generated by the subsymbols of $a^{(0)}$ over the ring of trigonometric polynomials equals the entire space. Then a consequence of the Quillen-Suslin Theorem is that we can complete the row $(A^{(0)}_{\gamma_0^*}, \ldots, A^{(0)}_{\gamma_{m-1}^*})$ to a matrix $\mathbf{A}^\top$ of trigonometric polynomials, which is invertible and whose inverse $\mathbf{A}^{-\top}$ also consists of trigonometric polynomials. Moreover, one can complete the matrix such that the first column of $\mathbf{A}^{-\top}$ equals $\frac{1}{m}(\overline{B^{(0)}_{\gamma_0^*}, \ldots, B^{(0)}_{\gamma_{m-1}^*}})^\top$, see also [Par95, JRS99] for details. With the choice $\overline{\mathbf{B}} := m\mathbf{A}^{-\top}$, we obtain

$$\mathbf{A}^\top \overline{\mathbf{B}} = m\mathcal{I}_m.$$

By applying (1.40), the subsymbols in $\mathbf{A}$ and $\mathbf{B}$ provide symbols $a^{(\mu)}$ and $b^{(\mu)}$, $\mu = 1, \ldots, m-1$. Hence, we finally found all of the necessary symbols to apply Theorem 1.1.11.

The original proof of the Quillen-Suslin Theorem is nonconstructive, but there are extension algorithms by means of Gröbner bases, which can be applied by some computer algebra software, see for example [Par95]. However, if there are other extension methods available, one would avoid the Gröbner approach since it can generally not guarantee for solutions with small masks.

In case $m = 2$, the extension problem is already solved to the full extent. The solution is summarized in the following example, which can be verified by completing the polyphase matrices and reconstructing the symbols from the subsymbols of the polyphase matrices.

Example 1.1.19. Let $m = 2$. Given a symbol $a^{(0)}$ and a dual symbol $b^{(0)}$, then (1.38) holds for the symbols

$$a^{(1)}(\xi) := e^{-2\pi i \gamma_1^* \cdot \xi} \overline{a^{(0)}(\xi + \gamma_1)},$$
$$b^{(1)}(\xi) := e^{-2\pi i \gamma_1^* \cdot \xi} \overline{b^{(0)}(\xi + \gamma_1)}.$$

The above choice is unique up to multiplication by $te^{-2\pi i (Ml) \cdot \xi}$, with $|t| = 1$ and $l \in \mathbb{Z}^d$. This affects the wavelets only by an l-shift and multiplication by t.

Note that Example 1.1.19 justifies the choice of $a^{(1)}$ in Examples 1.1.12, 1.1.13, and 1.1.14. The case $m > 2$ is generally much more involved. Nevertheless, for a large class of symbols, it is explicitly solvable as we shall explain in the following.

In Subsection 1.2.2, within the context of the fast wavelet transform, we discuss the importance of a *fundamental* refinable function, i.e., φ is continuous and

$$\varphi(k) = \delta_{0,k}, \quad \text{for all } k \in \mathbb{Z}^d.$$

By addressing the matrix completion, we focus on symbols a, whose refinable function is fundamental. It turns out a must then be *interpolatory*, i.e.,

$$\sum_{\gamma \in \Gamma_M} a(\xi + \gamma) = 1, \quad \text{for all } \xi \in \mathbb{R}^d. \tag{1.48}$$

The reverse implication holds if we assume φ to be stable and continuous, see [LLS97]. For an interpolatory symbol a, the results in [JRS99] provide a construction algorithm for the polyphase matrix completion. In the following lemma, we present the symbols according to the subsymbols of the polyphase matrices.

Lemma 1.1.20. *Given $a^{(0)}$ interpolatory and a dual symbol $b^{(0)}$. Then by the choice*

$$a^{(\mu)}(\xi) = e^{-2\pi i \gamma_\mu^* \cdot \xi} \sum_{\nu=1}^{m-1} \left(a^{(0)}(\xi + \gamma_\nu) - a^{(0)}(\xi) e^{-2\pi i \gamma_\mu^* \cdot \gamma_\nu} \right) \overline{b^{(0)}(\xi + \gamma_\nu)}, \tag{1.49}$$

$$b^{(\mu)}(\xi) = \frac{1}{m} e^{-2\pi i \gamma_\mu^* \cdot \xi} \sum_{\nu=1}^{m-1} \left(1 - e^{-2\pi i \gamma_\mu^* \cdot \gamma_\nu} \right) \overline{a^{(0)}(\xi + \gamma_\nu)}, \tag{1.50}$$

for $\mu = 1, \dots, m-1$, the matrix equality (1.38) *holds.*

Remark 1.1.21. As mentioned above, Lemma 1.1.20 is a consequence of the ideas in [JRS99]. Moreover, it is a special case of our much more general results of Corollary 4.2.6 in Chapter 4.

In the next example, we address a fundamental box spline:

Example 1.1.22. Given $M = 2\mathcal{I}_2$ and

$$Y_3 := \begin{pmatrix} 1 & 0 & 1 \\ 0 & 1 & 1 \end{pmatrix},$$

the matrix Y_3 is unimodular and the box spline symbol a_{Y_3} is interpolatory. Thus, the associated box spline φ_{Y_3} is fundamental. According to Lemma 1.1.17, φ_{Y_3} has globally linear independent integer shifts. Hence, there exists a dual symbol b, and Lemma 1.1.20 yields a collection of symbols such that (1.31) holds.

1.2 Desirable Properties

1.2.1 The Approximation Order

Given a Riesz basis $\{f_\kappa : \kappa \in \mathcal{K}\}$ for a Hilbert space $\mathcal{H}$, each $f \in \mathcal{H}$ has a series expansion

$$f = \sum_{\kappa \in \mathcal{K}} c_\kappa f_\kappa, \quad (c_\kappa)_{\kappa \in \mathcal{K}} \in \ell_2(\mathcal{K}).$$

From a computational point of view, it is necessary to replace the exact expansion by some approximation of f. General approximation theory is divided into linear and nonlinear methods. In Chapter 6, we consider nonlinear approximation, and we address the linear counterpart in the following.

Linear approximation centers around the approximation of $f \in \mathcal{H}$ from a given sequence $(\mathcal{V}_j)_{j \in \mathbb{Z}}$ of increasing linear subspaces of $\mathcal{H}$. The index j is often called a specific level or resolution. The error of best approximation of f from $(\mathcal{V}_j)_{j \in \mathbb{Z}}$ is expressed in the term $\operatorname{dist}(f, \mathcal{V}_j)_{\mathcal{H}}$. In order to approximate f at level $j \in \mathbb{Z}$ by a practical algorithm, we require a collection of linear mappings $Q_j : \mathcal{H} \to \mathcal{V}_j$, which represents a specific approximation. Since Q_j maps into $\mathcal{V}_j$, we obviously have

$$\operatorname{dist}(f, \mathcal{V}_j)_{\mathcal{H}} \leq \|f - Q_j f\|, \quad \text{for all } f \in \mathcal{H}.$$

To realize the best approximation, one has to find an approximation such that

$$\|f - Q_j f\| \sim \operatorname{dist}(f, \mathcal{V}_j)_{\mathcal{H}}.$$

In other words, $\operatorname{dist}(f, \mathcal{V}_j)_{\mathcal{H}}$ is a benchmark for each specific approximation.

Given an isotropic dilation matrix M, let us specify linear approximation with respect to a multiresolution analysis $(V_j)_{j\in\mathbb{Z}}$. It constitutes a sequence of increasing subspaces in $L_2(\mathbb{R}^d)$, and the following definition addresses the rate of best approximation, in which ρ denotes the modulus of the eigenvalues of M, see Appendix A.1 for the Sobolev space $W^s(L_2(\mathbb{R}^d))$:

Definition 1.2.1. Let M be isotropic. We say a multiresolution analysis $(V_j)_{j\in\mathbb{Z}}$ provides *approximation order* s if

$$\operatorname{dist}(f, V_j)_{L_2} \lesssim \rho^{-js}, \quad \text{for all } f \in W^s(L_2(\mathbb{R}^d)),$$

where the constant may depend on f.

In order to choose a specific approximation, let

$$X(\{\psi^{(1)}, \ldots, \psi^{(n)}\}), \quad X(\{\widetilde{\psi}^{(1)}, \ldots, \widetilde{\psi}^{(n)}\})$$

be a pair of biorthogonal wavelet bases, whose primal refinable function is the generator of the multiresolution analysis under consideration. In [DHRS03], the linear mapping

$$Q_j : L_2(\mathbb{R}^d) \to V_j, \quad f \mapsto \sum_{\substack{\mu=1,\ldots,n \\ j'<j,k\in\mathbb{Z}^d}} \left\langle f, \widetilde{\psi}^{(\mu)}_{j',k} \right\rangle \psi^{(\mu)}_{j',k} \tag{1.51}$$

is called the *truncated representation.* In order to evaluate the quality of its approximation, we consider the following definition:

Definition 1.2.2. Let M be isotropic. We say a pair of biorthogonal wavelet bases provides *approximation order* s if the truncated representation Q_j satisfies

$$\|f - Q_j f\|_{L_2} \lesssim \rho^{-js}, \quad \text{for all } f \in W^s(L_2(\mathbb{R}^d)),$$

where the constant may depend on f.

In the following, we shall verify that the truncated representations realizes the best approximation. Let φ and $\widetilde{\varphi}$ be the underlying primal and dual refinable functions, respectively. The operator

$$P_j : L_2(\mathbb{R}^d) \to V_j, \quad f \mapsto \sum_{k\in\mathbb{Z}^d} \langle f, \widetilde{\varphi}_{j,k} \rangle \varphi_{j,k}. \tag{1.52}$$

is a projection on V_j since the refinable functions are biorthogonal to each other. According to the results in [Lin05], P_j realizes the best approximation, i.e.,

$$\|f - P_j f\|_{L_2} \sim \operatorname{dist}(f, V_j)_{L_2}. \tag{1.53}$$

By applying the biorthogonality relations (1.19) and the decomposition (1.20), we obtain $Q_j = P_j$. Thus, the approximation order of a pair of biorthogonal wavelet bases equals the approximation order of the underlying primal multiresolution analysis.

In the sequel, we discuss properties of refinable functions and wavelets, which promote the approximation order. The ability of the underlying primal refinable function to reproduce polynomials plays a major role. Let us denote the space of all polynomials up to total degree less than s by Π_{s-1}. We say that a compactly supported distribution φ *reproduces polynomials up to order* s if

$$\Pi_{s-1} \subset S(\varphi),$$

where

$$S(\varphi) = \left\{ \sum_{k\in\mathbb{Z}^d} \lambda_k \varphi(\cdot - k) : (\lambda_k)_{k\in\mathbb{Z}^d} \in \ell(\mathbb{Z}^d) \right\}$$

denotes the *principal shift invariant space* spanned by φ. Note that $S(\varphi)$ makes sense as a subspace of $\mathcal{S}'(\mathbb{R}^d)$. According to the results in [Jia98], the approximation order of a multiresolution analysis is determined by its generator's reproduction of polynomials:

Theorem 1.2.3. *Given a dilation matrix $M = h\mathcal{I}_d$, $h \in \mathbb{N}$, let φ be a compactly supported, continuous generator of a multiresolution analysis with finitely supported mask and $\widehat{\varphi}(0) \neq 0$. Then φ reproduces polynomials up to order s iff its multiresolution analysis provides approximation order s.*

In [Lin05], Lindemann generalized one direction of Theorem 1.2.3 regarding isotropic dilation matrices:

Theorem 1.2.4. *Given an isotropic dilation matrix M, let φ be a compactly supported, continuous generator of a multiresolution analysis with finitely supported mask and $\widehat{\varphi}(0) \neq 0$. If φ reproduces polynomials up to order s, then its multiresolution analysis provides approximation order s.*

At least for idempotent dilation matrices, the reverse implication also holds, see Appendix A.2 for the proof of the following corollary, which extends Theorem 1.2.3:

Corollary 1.2.5. *Given an idempotent dilation matrix M, let φ be a compactly supported, continuous generator of a multiresolution analysis with finitely supported mask and $\widehat{\varphi}(0) \neq 0$. Then φ reproduces polynomials up to order s iff its multiresolution analysis provides approximation order s.*

Thus, the reproduction of polynomials determines the approximation order. According to the results of Jia in [Jia98], smoothness is a sufficient condition:

Theorem 1.2.6. *Given an isotropic dilation matrix M, for $s \in \mathbb{N}_0$, let φ be a compactly supported refinable function with finitely supported mask and $\widehat{\varphi}(0) \neq 0$. If φ is contained in $W^s(L_1(\mathbb{R}^d))$, then it reproduces polynomials up to order $s+1$.*

In general, we choose a symbol, and then φ is defined by (1.25). Fortunately, the reproduction of polynomials can be expressed in terms of its symbol. We say a symbol a satisfies the *sum rules of order* s if

$$\partial^\alpha a(\gamma) = 0, \quad \text{for all } \gamma \in \Gamma_M \setminus \{0\},\ |\alpha| < s. \tag{1.54}$$

According to [Jia98], the following dependencies hold:

Theorem 1.2.7. *If a symbol a, $a(0) = 1$, satisfies the sum rules of order s, then φ reproduces polynomials up to order s. If $\widehat{\varphi}$ has no $\mathbb{Z}^d$-periodical zeros on Γ_M, then also the reverse implication holds.*

A refinable function φ is stable iff its Fourier transform has no $\mathbb{Z}^d$-periodical zeros on $\mathbb{R}^d$, cf. [JM90]. Thus, for stable φ, the condition on $\widehat{\varphi}$ in Theorem 1.2.7 is satisfied.

Remark 1.2.8. According to Theorems 1.2.7 and 1.2.4 as well as the equivalence (1.53) with $Q_j = P_j$, the sum rule order of a implies approximation order of the associated biorthogonal wavelet bases.

Finally, in Section 5.2, we extend the concepts of multiresolution analysis and biorthogonal wavelet bases from $L_2(\mathbb{R}^d)$ to $L_p(\mathbb{R}^d)$ spaces with $1 < p < \infty$. Then one replaces $W^s(L_2(\mathbb{R}^d))$ by $W^s(L_p(\mathbb{R}^d))$, and best approximation in $L_p(\mathbb{R}^d)$ involves

$$\operatorname{dist}(f, V_j)_{L_p}.$$

It turns out that the results in $L_2(\mathbb{R}^d)$ presented so far essentially hold in $L_p(\mathbb{R}^d)$ as well, see [Lin05] and Chapter 5 in the present thesis.

1.2.2 Fast Wavelet Transform

Let φ and $\widetilde{\varphi}$ be compactly supported biorthogonal generators of two multiresolution analyses $(V_j)_{j\in\mathbb{Z}}$ and $(\widetilde{V}_j)_{j\in\mathbb{Z}}$, respectively. Then each $f \in V_0$ has the expansion

$$f = \sum_{k\in\mathbb{Z}^d} \langle f, \widetilde{\varphi}_{0,k}\rangle \, \varphi_{0,k}. \tag{1.55}$$

Given an associated pair $X(\{\psi^{(1)}, \dots, \psi^{(m-1)}\})$ and $X(\{\widetilde{\psi}^{(1)}, \dots, \widetilde{\psi}^{(m-1)}\})$ of compactly supported biorthogonal wavelet bases, let us fix some $j_0 \in \mathbb{Z}$ with $j_0 < 0$. Using the notation of Section 1.1.3, the complementary spaces $(W_j)_{j\in\mathbb{Z}}$ provide the decomposition

$$V_0 = V_{j_0} \oplus \bigoplus_{j=j_0}^{-1} W_j.$$

The biorthogonality relations (1.15) and (1.19) yield

$$f = \sum_{k\in\mathbb{Z}^d} \langle f, \widetilde{\varphi}_{j_0,k}\rangle \, \varphi_{j_0,k} + \sum_{j=j_0}^{-1} \sum_{\mu=1}^{m-1} \sum_{k\in\mathbb{Z}^d} \left\langle f, \widetilde{\psi}^{(\mu)}_{j,k}\right\rangle \psi^{(\mu)}_{j,k}. \tag{1.56}$$

Given the coefficients of f in V_0, i.e.,

$$H_0^{(0)}(k) := \langle f, \widetilde{\varphi}_{0,k}\rangle\,, \quad k \in \mathbb{Z}^d, \tag{1.57}$$

the fast wavelet transform computes the coefficients of the decomposition (1.56). Moreover, once derived the coefficients in (1.56), the transform can reconstruct the coefficients in (1.55), cf. Algorithm 1, Figure 1.4, and see Appendix A.3 for the arising notation. Hence, the fast wavelet transform provides a tool to switch between the two representations.

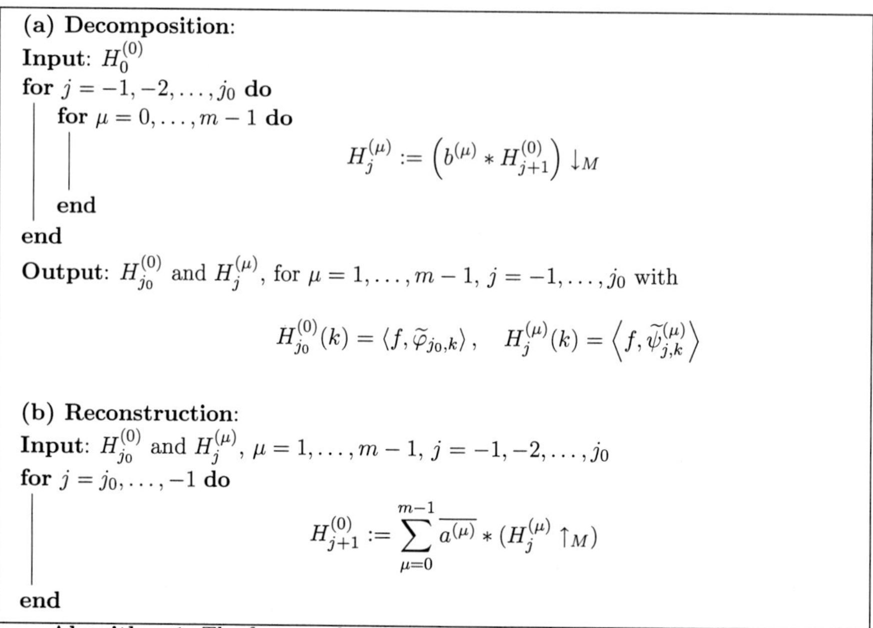

(a) Decomposition:
Input: $H_0^{(0)}$
for $j = -1, -2, \ldots, j_0$ **do**
  for $\mu = 0, \ldots, m-1$ **do**

$$H_j^{(\mu)} := \left(b^{(\mu)} * H_{j+1}^{(0)}\right) \downarrow_M$$

  end
end
Output: $H_{j_0}^{(0)}$ and $H_j^{(\mu)}$, for $\mu = 1, \ldots, m-1$, $j = -1, \ldots, j_0$ with

$$H_{j_0}^{(0)}(k) = \langle f, \widetilde{\varphi}_{j_0,k} \rangle, \quad H_j^{(\mu)}(k) = \left\langle f, \widetilde{\psi}_{j,k}^{(\mu)} \right\rangle$$

(b) Reconstruction:
Input: $H_{j_0}^{(0)}$ and $H_j^{(\mu)}$, $\mu = 1, \ldots, m-1$, $j = -1, -2, \ldots, j_0$
for $j = j_0, \ldots, -1$ **do**

$$H_{j+1}^{(0)} := \sum_{\mu=0}^{m-1} \overline{a^{(\mu)}} * (H_j^{(\mu)} \uparrow_M)$$

end

Algorithm 1: The fast wavelet transform, see Appendix A.3 for notation

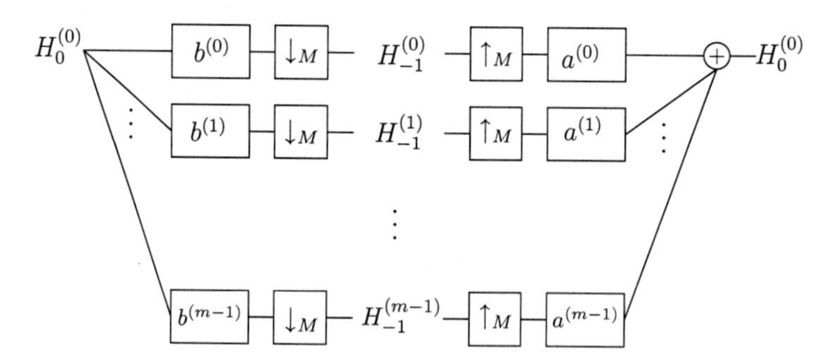

Figure 1.4: The filter bank scheme of the fast wavelet transform

Next, in view of the fast wavelet transform, we discuss desirable properties of wavelets and underlying refinable functions.

Fundamental

The computation of the coefficients in (1.57) is often expensive and sometimes even impossible. A quite common method is to replace the inner products by sample values of f. However, Strang and Nguyen call this approach

"...the wavelet crime.",

see [SN96]. Nevertheless, if φ is fundamental, then the expansion (1.55) implies

$$f(k) = \langle f, \widetilde{\varphi}_{0,k} \rangle, \quad k \in \mathbb{Z}^d.$$

Hence, the coefficients are indeed sample values. In the fundamental setting, the wavelet crime is legal.

Small Support, Smoothness

Since all functions φ, $\widetilde{\varphi}$, and $\psi^{(\mu)}$, $\widetilde{\psi}^{(\mu)}$, $\mu = 1, \dots, m-1$, are compactly supported, the coefficients

$$\langle f, \widetilde{\varphi}_{j,k} \rangle, \quad \left\langle f, \widetilde{\psi}^{(\mu)}_{j,k} \right\rangle$$

contain local information about f, and the smaller the support sizes, the better local properties can be read off the coefficients. The support of wavelets and refinable functions can be estimated by the support sizes of their underlying masks, see [GM92]: let the symbol a, $a(0) = 1$, generate the refinable function φ, then

$$\operatorname{supp}(\varphi) \subset \sum_{j=1}^{\infty} M^{-j} \operatorname{supp}\left((a_k)_{k \in \mathbb{Z}^d}\right). \tag{1.58}$$

In Examples 1.1.12, 1.1.13, and 1.1.14, we even have equality for (1.58). Thus, small mask sizes are important because they bound the supports of the underlying refinable function and wavelets. Moreover, small masks are desirable for their own since they provide small filter lengths in the transform, which reduces the complexity of the filter bank convolutions.

In general, the signal f inherits at least to a certain extent some smoothness. It is reasonable that primal wavelets and refinable function should also share this smoothness. Although theoretical results in wavelet analysis do not require smooth dual wavelets, numerical experiments in image and signal processing provide better results if primal and dual wavelets have similar smoothness. This observation can be justified as follows: if primal and dual wavelets have similar support size and similar smoothness, then decomposition and reconstruction resemble the orthonormal setting, which is the most economic. Thus, localization does not only involve small supports, but also similar smoothness of primal and dual wavelets. If primal and dual refinable functions coincide, then each wavelet inherits exactly the same smoothness, which is optimal.

Reproduction of Polynomials, Vanishing Moments

Let φ reproduce polynomials up to order s. Then the biorthogonality relations (1.18) yield that all dual wavelets $\widetilde{\psi}^{(\mu)}$, $\mu = 1, \dots, m-1$, have s *vanishing moments*, i.e.,

$$\int_{\mathbb{R}^d} x^{\alpha} \widetilde{\psi}^{(\mu)}(x) dx = 0, \quad \text{for all } |\alpha| < s. \tag{1.59}$$

Hence, the wavelet coefficient $\left\langle f, \widetilde{\psi}_{j,k}^{(\mu)} \right\rangle$ measures how much f changes in a neighborhood of $M^{-j}\left(\operatorname{supp}(\widetilde{\psi}^{(\mu)}) + k\right)$. For smooth f, the coefficient is small. If f has some kind of singularity, then it may become large. In other words, the wavelet coefficients detect the details of f. Since φ reproduces polynomials up to order s, the smooth parts of f are covered by the shifts of φ. Hence, the fast wavelet transform separates the smooth parts of the signal from its details. Finally, a high order of polynomial reproduction and a high number of vanishing moments promote this separation.

Real-Valued, Symmetric, Nonseparable

In most applications, the signal f is real-valued. Hence, we prefer real-valued refinable functions and wavelets. Moreover, symmetric and nonseparable wavelets are claimed to provide better results in image and signal analysis. For instance, engineers often prefer *linear phase* filters, i.e., filters that are symmetric about a point, cf. [Mal89]. If the refinable function's filter is linear phase, then the refinable function itself is symmetric about a point. Similar results hold for the associated wavelets as well as for axis symmetries, see [Han04] and Section 2.3.2 for details.

Small Number of Wavelets

The complexity of the fast wavelet transform is directly influenced by the number of wavelets. As already mentioned in Section 1.1.3, this number equals $m-1$. Moreover, m has the interpretation of a scaling difference, i.e., the changeover from V_j to V_{j+1}. Its minimization allows for a desirable finer analysis of signals and images, see for instance [MPMK98]. Since the dilation matrix has to be expanding, $m=2$ minimizes the complexity and the scaling difference. Fortunately, such dilation matrices exist in all dimensions, see for instance (2.27) and (2.28).

1.2.3 The Characterization of Smoothness Classes

In many problems of applied mathematics, one has to address smoothness spaces. Such function spaces play key roles in the treatment of variational problems, partial differential equations, as well as operator equations. In order to derive solutions from practical algorithms, we have to discretize the problems. Wavelet bases provide a valuable tool as we shall describe in the sequel, see Chapter 5 for a more detailed discussion about this topic.

Biorthogonal wavelets characterize many smoothness classes, i.e., they provide series expansions and the smoothness norm is equivalent to a weighted sequence norm of wavelet coefficients, see [Lin05] and Chapter 5 for details. Hence, the result is an efficient discretization of the original problem. However, the characterization requires that the primal refinable function has sufficient smoothness as well as polynomial reproduction, and the dual wavelets have a sufficient number of vanishing moments.

According to Theorem 1.2.6, smoothness of the primal refinable function implies reproduction of polynomials, which yields vanishing moments of the dual wavelets. Thus, the assumptions for the characterization reduce to the smoothness condition. In fact, smoothness can be considered as one of the strongest requirements in compactly supported pairs of biorthogonal wavelet bases.

1.3 Restrictions

Let φ and $\widetilde{\varphi}$ be biorthogonal generators of two multiresolution analyses $(V_j)_{j\in\mathbb{Z}}$ and $(\widetilde{V}_j)_{j\in\mathbb{Z}}$, respectively. Given an associated pair of biorthogonal wavelet bases

$$X(\{\psi^{(1)},\dots,\psi^{(m-1)}\}),\quad X(\{\widetilde{\psi}^{(1)},\dots,\widetilde{\psi}^{(m-1)}\}),$$

the discussion of Section 1.2 leads to the following wish list:

- smoothness,
- small support size,
- reproduction of polynomials up to a high order,
- high number of vanishing moments,
- small number of wavelets,
- fundamental refinable functions,
- symmetric,
- real-valued,
- nonseparable,
- primal and dual refinable functions coincide.

Many items listed above depend on each other, and these dependencies entail some restrictions, which are the topics for the remainder of this chapter. We proceed as follows: first, we mention some overall restrictions. They bother all wavelet bases constructions. Then we go into detail, and we divide our discussion into restrictions of orthonormal wavelets as well as restrictions of biorthogonal wavelets. This division may guarantee a detailed discussion.

Overall Restrictions

The support sizes of refinable functions and wavelets compete with their smoothness. In other words, given a certain support size, the smoothness is bounded from above. Unfortunately, the bound is often quite low, which means a serious limitation. The reproduction of polynomials and support sizes are also competing, but the arising restrictions are less crucial.

Next, we address relations between the number of wavelets and fundamental refinable functions. In order to minimize the number of wavelets, one requires a dilation matrix with $m = 2$. Then refinable functions suffer from the following structural restriction, see Appendix A.2 for its proof:

Lemma 1.3.1. *Given a pair of compactly supported biorthogonal wavelet bases, let their associated refinable functions φ and $\widetilde{\varphi}$ be generated by symbols a and b, respectively, with $a(0) = b(0) = 1$. If both refinable functions are fundamental, then $m > 2$.*

The effects of Lemma 1.3.1 are considered in the following specific situations.

Orthonormal Wavelets

Let M be a dilation matrix with $m = 2$. Then Lemma 1.3.1 is deflating for orthogonal wavelets. It tells us that the underlying refinable function is not fundamental, see also [Dau92] for the univariate statement. The situation is even worse. Let $X(\{\psi\})$ be a compactly supported real-valued orthonormal wavelet basis. If ψ is continuous, then it is not symmetric about a point, cf. [Han04]. Hence, the filter is not linear phase.

In general, it is impossible to adapt univariate constructions of orthonormal wavelet bases to the multivariate setting since they use a certain factorization technique that does not hold in higher dimensions. In order to circumvent this problem, one often derives multivariate wavelets from tensor products of dyadic univariate wavelets, see Example 1.1.15. On the one hand, they inherit the univariate drawbacks, such as missing symmetries or a nonfundamental refinable function. On the other hand, a d-dimensional tensor approach yields wavelet bases of $2^d - 1$ wavelets, i.e., the number increases exponentially, which means that the complexity of the fast wavelet transform becomes too large.

Thus, we require nonseparable multivariate wavelets. However, we already face serious difficulties in the bivariate setting. For the popular quincunx dilation matrix M_q, one times differentiable orthonormal wavelets with compact support could not yet be constructed.

One overcomes some of the restrictions and difficulties of orthonormal wavelets with the concept of pairs of biorthogonal wavelet bases. However, as we shall discuss next, they still provide some undesirable constraints:

Biorthogonal Wavelets

At first glance, Lemma 1.3.1 is harmless for biorthogonal wavelet bases. In order to consider the wavelet coefficients as sample values in the fast wavelet transform, we require a fundamental φ, but we do not need $\widetilde{\varphi}$ to be fundamental as well. Moreover, the concept of pairs of biorthogonal wavelet bases allows for symmetric wavelets, see [CDF92, CD93, Der99, JRS99] for successful constructions of compactly supported symmetric wavelets with a fundamental refinable function.

In order to minimize the number of wavelets, one must choose a dilation matrix with $m = 2$. If the primal refinable function is then fundamental, Lemma 1.3.1 prohibits the last point on our wish list, i.e., the dual refinable function does not coincide with the primal one. While the last mentioned limitation may not be crucial, the following limitation is definitely serious. One can construct compactly supported nonseparable biorthogonal wavelet bases such that the primal refinable function is smooth and fundamental with a small support. However, these strong primal properties are generally accompanied by a weak dual refinable function, i.e., the dual refinable function has either poor smoothness or a large support.

Finally, not only orthogonality but also biorthogonality is such a strong property that not all items on our wish list can be incorporated. We conclude this chapter with Table 1.1. It summarizes the main limitations of orthonormal and biorthogonal wavelet bases.

features	limitations
orthonormal $m = 2$ compact support	not fundamental
tensor approach compact support	
orthonormal $m = 2$ continuous compact support real-valued	no symmetry about a point
orthonormal quincunx compact support	not $\mathcal{C}^1(\mathbb{R}^2)$ so far
biorthogonal compact support smoothness	large support

Table 1.1: Restrictions of wavelet bases

Chapter 2

More Flexibility: Wavelet Bi-Frames

As mentioned in Section 1.3, pairs of biorthogonal wavelet bases do not allow for all desirable properties of wavelets. This chapter is dedicated to the introduction and discussion of a weaker concept, namely pairs of dual wavelet frames, called wavelet bi-frames for short. The concept generalizes pairs of biorthogonal wavelet bases. Although primal and dual wavelets are not required to satisfy any biorthogonality relations, they still provide some series expansion. Similar to biorthogonal bases, the coefficients can be derived by inner products with the dual wavelets. Contrary to bases, the frame concept allows for redundancy. Hence, the coefficients may not be unique.

In order to construct wavelet bi-frames, one generally adapts the multiresolution analysis approach in Chapter 1, and wavelets are obtained from underlying refinable functions. Contrary to the biorthogonal setting, wavelets are neither supposed to span any complementary spaces nor do they have to satisfy any geometrical relations. This also means that primal and dual refinable functions are no longer required to be biorthogonal, which provides much flexibility for their choice.

However, the weaker frame concept can yield some drawbacks. Contrary to biorthogonal wavelets, the wavelet bi-frame's approximation order can be less than the one of the underlying multiresolution analysis. In other words, it may happen that the wavelet bi-frame does not use the potential of the underlying multiresolution analysis to full capacity. Nevertheless, if the wavelet bi-frame has sufficient vanishing moments, then its approximation order reaches the one of the multiresolution analysis, and one overcomes this problem. In fact, the number of vanishing moments can be considered as a major quality critirion of a wavelet bi-frame.

First, following the textbook [Chr03], we introduce frames in Hilbert spaces. Then we apply the concept to wavelets, and we establish the so-called mixed extension principle. It provides a construction method for wavelet bi-frames, which is similar to the method in Theorem 1.1.11 addressing biorthogonal wavelets. Since primal and dual refinable functions are decoupled, the concept allows for the construction of primal and dual wavelets from one single refinable function, whose integer shifts are not orthogonal.

Wavelet bi-frames from the mixed extension principle provide a fast wavelet frame transform. The filter bank scheme is nearly identical to the one of the fast wavelet transform in Figure 1.4. It merely allows for more channels, i.e., $m-1$ is replaced by a possibly larger number n. However, the lack of the biorthogonality relations requires some additional attention to the interpretation of the computed sequences as we shall explain in Subsection 2.3.2. Then, we introduce several optimality criteria for wavelet bi-frames. They serve as benchmarks for the constructions in the following Chapters 3 and 4. We consider optimality with respect to approximation order, i.e., the approximation order of the wavelet bi-frame reaches the one of the underlying multiresolution analysis. Moreover, we derive optimality conditions regarding sum rules, smoothness, and the refinable function's mask size. Finally, we address symmetry and the optimal number of wavelets. We conclude this chapter with a summary of the established optimality criteria.

2.1 Wavelet Frames

2.1.1 Frames in Hilbert Spaces

Let $\mathcal{K}$ be a countable index set with some ordering. A collection $\{f_\kappa : \kappa \in \mathcal{K}\}$ in a Hilbert space $\mathcal{H}$ is called a *Bessel sequence* if there is a positive constant B such that

$$\left\|(\langle f, f_\kappa\rangle)_{\kappa\in\mathcal{K}}\right\|_{\ell_2}^2 \leq B\|f\|_{\mathcal{H}}^2, \quad \text{for all } f \in \mathcal{H}.$$

It turns out that $\{f_\kappa : \kappa \in \mathcal{K}\}$ is a Bessel sequence iff its *synthesis operator*

$$F : \ell_2(\mathcal{K}) \to \mathcal{H}, \quad (c_\kappa)_{\kappa\in\mathcal{K}} \mapsto \sum_{\kappa\in\mathcal{K}} c_\kappa f_\kappa,$$

is well-defined. Then

$$\sum_{\kappa\in\mathcal{K}} c_\kappa f_\kappa$$

converges unconditionally, and the Banach-Steinhaus Theorem yields that F is already bounded, see Section 3.2 in [Chr03] for details. In order to obtain a series expansion for all $f \in \mathcal{H}$, we need a surjective synthesis operator. This motivates the following definition.

Definition 2.1.1. Let $\mathcal{K}$ be a countable index set. A collection $\{f_\kappa : \kappa \in \mathcal{K}\}$ in a Hilbert space $\mathcal{H}$ is called a *frame* for $\mathcal{H}$ if there exist two positive constants A, B such that

$$A\|f\|_{\mathcal{H}}^2 \leq \|(\langle f, f_\kappa\rangle)_{\kappa\in\mathcal{K}}\|_{\ell_2}^2 \leq B\|f\|_{\mathcal{H}}^2, \quad \text{for all } f \in \mathcal{H}. \tag{2.1}$$

Then A and B are called the *lower* and *upper frame bounds*, respectively.

The collection $\{f_\kappa : \kappa \in \mathcal{K}\}$ is a frame iff its synthesis operator $F : \ell_2(\mathcal{K}) \to \mathcal{H}$ is well-defined and onto, see Section 5.5 in [Chr03]. Hence, each $f \in \mathcal{H}$ has a series expansion in the frame. This expansion is not required being unique, and a frame is called *overcomplete* if the synthesis operator has a nontrivial kernel.

In order to get closer to the frame concept, we present some examples. In finite-dimensional Hilber spaces, the characterization of frames is quite simple:

Example 2.1.2. If $\dim(\mathcal{H}) < \infty$, then $\{f_1, \ldots, f_n\}$ is a frame for $\mathcal{H}$ iff it spans $\mathcal{H}$.

Next, we address the relations between Riesz bases and frames.

Example 2.1.3. If $\{f_\kappa : \kappa \in \mathcal{K}\}$ is a Riesz basis for $\mathcal{H}$, then it is also a frame for $\mathcal{H}$, and the Riesz bounds coincide with the frame bounds. It is not overcomplete.

In view of Example 2.1.3, one may ask for the differences between Riesz bases and frames. The following theorem provides an answer, cf. Sections 3.4 and 5.5 in [Chr03].

Theorem 2.1.4. *Let $\{e_\kappa : \kappa \in \mathcal{K}\}$ be an orthonormal basis for $\mathcal{H}$ with* $\mathrm{card}(\mathcal{K}) = \infty$. *Then the following holds:*

$\{f_\kappa : \kappa \in \mathcal{K}\}$	$F : \ell_2(\mathcal{K}) \to \mathcal{H}$
Bessel sequence	well-defined
Frame	onto
Riesz basis	invertible
Orthonormal basis	unitary

Table 2.1: Concepts in terms of the synthesis operator

(a) *The Riesz bases for $\mathcal{H}$ are precisely the families $\{Ue_\kappa : \kappa \in \mathcal{K}\}$, where U is a bounded and invertible operator on $\mathcal{H}$.*

(b) *The frames for $\mathcal{H}$ are precisely the families $\{Ue_\kappa : \kappa \in \mathcal{K}\}$, where U is a bounded and surjective operator on $\mathcal{H}$.*

Table 2.1 provides a summary of the differences between Bessel sequences, frames, Riesz bases, and orthonormal bases in terms of their synthesis operators. Moreover, the following result yields a second description of the differences between Riesz bases and frames, see Section 6.1 in [Chr03]. Recall that a frame is called *exact* if it ceases to be a frame when an arbitrary element is removed.

Theorem 2.1.5. *Let $\{f_\kappa : \kappa \in \mathcal{K}\}$ be a frame. Then the following are equivalent:*

(i) *$\{f_\kappa : \kappa \in \mathcal{K}\}$ is a Riesz basis.*

(ii) *$\{f_\kappa : \kappa \in \mathcal{K}\}$ is an exact frame.*

Given a Bessel sequence $\{f_\kappa : \kappa \in \mathcal{K}\}$ for $\mathcal{H}$, the adjoint of the synthesis operator, i.e.,

$$F^* : \mathcal{H} \to \ell_2(\mathcal{K}), \quad f \mapsto (\langle f, f_\kappa \rangle)_{\kappa \in \mathcal{K}} \tag{2.2}$$

is called the *analysis operator*. The following simple lemma shows that biorthogonal frames are already Riesz bases. Since it is a combination of Theorems 3.6.3 and 6.1.1 in [Chr03], we present the proof in Appendix A.2 for the sake of completeness.

Lemma 2.1.6. *Given a frame $\{f_\kappa : \kappa \in \mathcal{K}\}$ for $\mathcal{H}$, let $\{\widetilde{f}_\kappa : \kappa \in \mathcal{K}\} \subset \mathcal{H}$ be a biorthogonal sequence, i.e.,*

$$\left\langle f_\kappa, \widetilde{f}_{\kappa'} \right\rangle = \delta_{\kappa,\kappa'}, \quad \textit{for all } \kappa, \kappa' \in \mathcal{K}.$$

Then both systems constitute a pair of biorthogonal Riesz bases for $\mathcal{H}$.

Given a frame $\{f_\kappa : \kappa \in \mathcal{K}\}$, the synthesis operator $F : \ell_2(\mathcal{K}) \to \mathcal{H}$ is onto. Hence, each $f \in \mathcal{H}$ has a series expansion. However, we have no tool so far to determine the coefficients. In the sequel, it turns out that there is a second frame, which computes the coefficients by inner products, and f can be expanded similar to the biorthogonal Riesz bases setting.

The operator

$$S := FF^* : \mathcal{H} \to \mathcal{H}, \quad f \mapsto \sum_{\kappa \in \mathcal{K}} \langle f, f_\kappa \rangle f_\kappa$$

is called the *frame operator*. It is positive and boundedly invertible, and the system $\{S^{-1}f_\kappa : \kappa \in \mathcal{K}\}$ is called the *canonical dual frame*, cf. Chapter 5 in [Chr03]. It is a frame, and it provides the expansion

$$f = \sum_{\kappa\in\mathcal{K}} \langle f, S^{-1}f_\kappa\rangle f_\kappa, \quad \text{for all } f \in \mathcal{H}. \tag{2.3}$$

Moreover, the canonical dual frame has the following minimality property. Given $f \in \mathcal{H}$, for all sequences $(c_\kappa)_{\kappa\in\mathcal{K}}$ such that

$$f = \sum_{\kappa\in\mathcal{K}} c_\kappa f_\kappa,$$

we have

$$\left\|(c_\kappa)_{\kappa\in\mathcal{K}}\right\|_{\ell_2}^2 = \left\|\left(\langle f, S^{-1}f_\kappa\rangle\right)_{\kappa\in\mathcal{K}}\right\|_{\ell_2}^2 + \left\|\left(\langle f, S^{-1}f_\kappa\rangle - c_\kappa\right)_{\kappa\in\mathcal{K}}\right\|_{\ell_2}^2. \tag{2.4}$$

Hence, the coefficients provided by the canonical dual frame have minimal ℓ_2-norm among all possible choices of coefficients.

Let us present a simple example of a frame for $\mathbb{R}^2$ and its canonical dual frame. It is not contained in [Chr03]:

Example 2.1.7. Let

$$f_1 := \begin{pmatrix}1\\0\end{pmatrix}, \quad f_2 := \begin{pmatrix}0\\1\end{pmatrix}, \quad f_3 := -\frac{1}{\sqrt{2}}\begin{pmatrix}1\\1\end{pmatrix},$$

then $\{f_1, f_2, f_3\}$ is a frame for $\mathbb{R}^2$ with bounds $A = 1$, $B = 2$. The frame operator and its inverse are given by

$$S = \frac{1}{2}\begin{pmatrix}3 & 1\\1 & 3\end{pmatrix}, \quad S^{-1} = \frac{1}{4}\begin{pmatrix}3 & -1\\-1 & 3\end{pmatrix}.$$

Hence, the canonical dual frame is

$$\left\{\frac{1}{4}\begin{pmatrix}3\\-1\end{pmatrix}, \frac{1}{4}\begin{pmatrix}-1\\3\end{pmatrix}, -\frac{1}{2\sqrt{2}}\begin{pmatrix}1\\1\end{pmatrix}\right\}. \tag{2.5}$$

See Figure 2.1 for a visualization of both frames in $\mathbb{R}^2$.

If we can choose $A = B$ in (2.1), then $\{f_\kappa : \kappa \in \mathcal{K}\}$ is called a *tight frame*. It yields $S = A \cdot \mathrm{id}_\mathcal{H}$, and the inversion of S is trivial, cf. Section 5.7 in [Chr03]. Thus, although the elements are not orthogonal in general, the tight frame yields an expansion like (1.2) by

$$f = \frac{1}{A}\sum_{\kappa\in\mathcal{K}} \langle f, f_\kappa\rangle f_\kappa.$$

In other words, tight frames generalize orthonormal bases. For instance, the uniform arrangement of three vectors in $\mathbb{R}^2$ provides a tight frame, cf. [Dau92]:

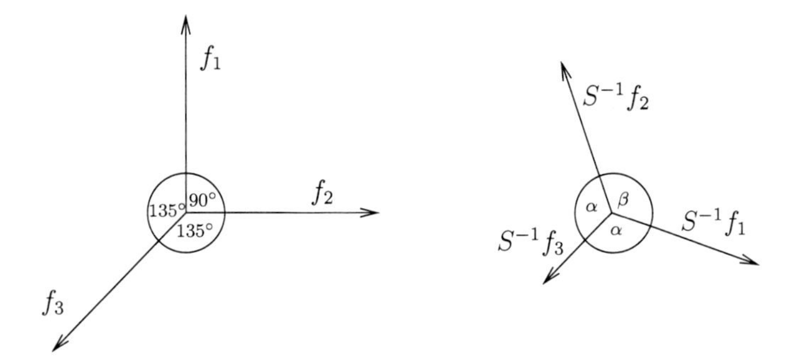

Figure 2.1: Primal frame and its canonical dual in Example 2.1.7. We have $\alpha \approx 116.57°$ and $\beta \approx 126.87°$.

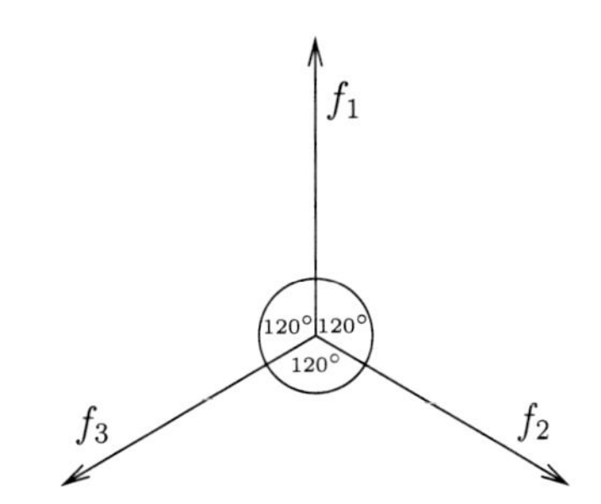

Figure 2.2: Tight frame in Example 2.1.8

Example 2.1.8. Let

$$f_1 := \begin{pmatrix} 0 \\ 1 \end{pmatrix}, \quad f_2 := -\frac{1}{2}\begin{pmatrix} \sqrt{3} \\ 1 \end{pmatrix}, \quad f_3 := \frac{1}{2}\begin{pmatrix} \sqrt{3} \\ -1 \end{pmatrix},$$

then $\{f_1, f_2, f_3\}$ is a tight frame for $\mathbb{R}^2$ with frame bounds $A = B = \frac{3}{2}$, see Figure 2.2.

In infinite-dimensional Hilbert spaces, there exist overcomplete tight frames, whose finite subsets are linearly independent, see [Chr03]:

Example 2.1.9. It is well-known that $\{e^{2\pi ikx}|_{(0,1)} : k \in \mathbb{Z}\}$ is an orthonormal basis for $L_2(0,1)$. Let $J \subset (0,1)$ be a pure subinterval. Then $\{e^{2\pi ikx}|_J : k \in \mathbb{Z}\}$ is an overcomplete tight frame for $L_2(J)$. It is noteworthy that each finite subset is linearly independent, cf. Section 1.6 in [Chr03].

In the sequel, we consider frames, which inherit a certain structure. First, we address Gabor frames. They are extensively studied in the context of time frequency analysis, see the textbooks [FS03, Grö01]. For $a, b \in \mathbb{R}$, let

$$\begin{aligned} T_a : L_2(\mathbb{R}) \to L_2(\mathbb{R}), \quad (T_a f)(x) &:= f(x-a), \\ E_b : L_2(\mathbb{R}) \to L_2(\mathbb{R}), \quad (E_b f)(x) &:= e^{2\pi ibx} f(x), \end{aligned}$$

denote translation and modulation operators, respectively. For $g \in L_2(\mathbb{R})$, the collection $\{E_{mb}T_{an}g : m, n \in \mathbb{Z}\}$ is called a *Gabor frame* if it is a frame for $L_2(\mathbb{R})$. The following example is borrowed from Section 8.6 in [Chr03]:

Example 2.1.10. Let $a, b > 0$ and $g(x) := e^{-x^2}$. Then $\{E_{mb}T_{an}g : m, n \in \mathbb{Z}\}$ is a Gabor frame iff $ab < 1$.

Given $g \in L_2(\mathbb{R})$ and $a, b > 0$, let $\{E_{mb}T_{an}g : m, n \in \mathbb{Z}\}$ be a Gabor frame. Then the frame operator S commutes with the translation and modulation operators, i.e.,

$$SE_{mb}T_{na} = E_{mb}T_{na}S, \quad \text{for all } m, n \in \mathbb{Z}.$$

This implies $E_{mb}T_{na} = S^{-1}E_{mb}T_{na}S$, and we finally obtain

$$E_{mb}T_{na}S^{-1} = S^{-1}E_{mb}T_{na}, \quad \text{for all } m, n \in \mathbb{Z}.$$

Thus, the canonical dual frame is

$$\{E_{mb}T_{an}S^{-1}g : m, n \in \mathbb{Z}\}. \tag{2.6}$$

This provides a comfortable feature of Gabor frames: their canonical dual is also a Gabor frame.

Next, we address the wavelet structure. Given $\psi^{(1)}, \ldots, \psi^{(n)} \in L_2(\mathbb{R}^d)$, we call $X(\{\psi^{(1)}, \ldots, \psi^{(n)}\})$ a *wavelet frame* if it constitutes a frame for $L_2(\mathbb{R}^d)$. Unfortunately, dilation and translation generally do not commute with the frame operator. Hence, its canonical dual may not have the wavelet structure. The following Subsection 2.1.2 provides an alternative, which circumvents this inconvenience.

2.1.2 Bi-Frames

The frame concept allows for redundancy. Thus, there can be further expansions similar to (2.3). Given the frame $\{f_1, f_2, f_3\}$ in Example 2.1.7, then $\{f_1, f_2\}$ is already an orthonormal basis for $\mathbb{R}^2$. Hence,

$$\{\widetilde{f}_1, \widetilde{f}_2, \widetilde{f}_3\} := \{f_1, f_2, 0\} \tag{2.7}$$

is a frame which provides the expansion

$$f = \sum_{\kappa=1}^{2} \left\langle f, \widetilde{f}_\kappa \right\rangle f_\kappa, \quad \text{for all } f \in \mathbb{R}^2.$$

Thus, the choice (2.7) essentially yields the usual orthonormal expansion. Hence, non-canonical choices can provide some features.

Given a frame of wavelets, we have already discussed that its canonical dual frame may not have the wavelet structure as well. Nevertheless, the canonical dual can possibly be replaced by an alternative dual wavelet frame. This motivates the following definition.

Definition 2.1.11. Two Bessel sequences $\{f_\kappa : \kappa \in \mathcal{K}\}$ and $\{\widetilde{f}_\kappa : \kappa \in \mathcal{K}\}$ for $\mathcal{H}$ are called a *pair of dual frames* (or a *bi-frame*) if the expansion

$$f = \sum_{\kappa \in \mathcal{K}} \left\langle f, \widetilde{f}_\kappa \right\rangle f_\kappa \tag{2.8}$$

holds for every $f \in \mathcal{H}$.

Given a bi-frame, (2.8) means $\mathrm{id}_\mathcal{H} = F\widetilde{F}^*$, where F and $\widetilde{F}$ are the associated synthesis operators. Then F is surjective, and $\widetilde{F}^*$ is injective, which yields that $\widetilde{F}$ is surjective. Hence, both pairs of a bi-frame are actually frames for $\mathcal{H}$, see also Chapter 5 in [Chr03].

Next, we apply the bi-frame concept to wavelets.

Definition 2.1.12. Given $\psi^{(\mu)}, \widetilde{\psi}^{(\mu)} \in L_2(\mathbb{R}^d)$, $\mu = 1, \ldots, n$, then

$$X(\{\psi^{(1)}, \ldots, \psi^{(n)}\}), \quad X(\{\widetilde{\psi}^{(1)}, \ldots, \widetilde{\psi}^{(n)}\})$$

are called a *wavelet bi-frame* if they constitute a pair of dual frames in $L_2(\mathbb{R}^d)$.

For a wavelet bi-frame $X(\{\psi^{(1)}, \ldots, \psi^{(n)}\})$, $X(\{\widetilde{\psi}^{(1)}, \ldots, \widetilde{\psi}^{(n)}\})$, the usual bi-frame expansion (2.8) reads as

$$f = \sum_{\mu=1}^{n} \sum_{j \in \mathbb{Z}} \sum_{k \in \mathbb{Z}^d} \left\langle f, \widetilde{\psi}^{(\mu)}_{j,k} \right\rangle \psi^{(\mu)}_{j,k}, \quad \text{for } f \in L_2(\mathbb{R}^d). \tag{2.9}$$

Hence, they combine the advantages of Gabor frames and tight frames: similar to the Gabor frames, primal and dual frame share the same structure. As with tight frames, one does not require the inversion of the frame operator.

2.2 The Mixed Extension Principle

The wavelet construction of Section 1.1.3 is based on a refinable function φ generating a multiresolution analysis $(V_j)_{j\in\mathbb{Z}}$. The closed linear span of the integer shifts of the wavelets is an algebraic complement of V_0 in V_1. Thus, the wavelets are contained in V_1. Then they can be represented by linear combinations of $\varphi(Mx-k)$, $k\in\mathbb{Z}^d$.

In the wavelet frame setting, we still use this approach, but we can weaken the concept of multiresolution analysis in Definition 1.1.9. We replace (M-4) by the weaker condition

(M-4') there is a function $\varphi\in V_0$ such that V_0 is the closed linear span of its integer shifts.

In other words, we can skip the stability assumption on the generator φ. According to [dBDR93], each compactly supported refinable function $\varphi\in L_2(\mathbb{R}^d)$ with $\widehat{\varphi}(0)\neq 0$ generates a multiresolution analysis with (M-4) replaced by (M-4'). Thus, we do not require any stability.

As usual, we start with two compactly supported refinable functions φ and $\widetilde{\varphi}$ contained in $L_2(\mathbb{R}^d)$. They are implicitly given by two symbols $a^{(0)}$ and $b^{(0)}$. The wavelets are defined by

$$\psi^{(\mu)}(x):=\sum_{k\in\mathbb{Z}^d}a_k^{(\mu)}\varphi(Mx-k),\qquad \widetilde{\psi}^{(\mu)}(x):=\sum_{k\in\mathbb{Z}^d}b_k^{(\mu)}\widetilde{\varphi}(Mx-k),\tag{2.10}$$

where $a^{(\mu)},b^{(\mu)}$ are additional symbols, for $\mu=1,\dots,n$. Note the only difference to (1.27) and (1.28): we allow for $n\geq m-1$. Since the frame concept abandons biorthogonality as well as stability of the generators, neither the relations (1.15) and (1.16) nor the decompositions (1.14) are required to hold.

In the sequel, we discuss conditions which ensure that

$$X(\{\psi^{(1)},\dots,\psi^{(n)}\}),\quad X(\{\widetilde{\psi}^{(1)},\dots,\widetilde{\psi}^{(n)}\})$$

constitute a wavelet bi-frame. We say the symbol family

$$\{(a^{(\mu)},b^{(\mu)}):\mu=0,\dots,n\}$$

satisfies *condition (I)* if the following holds:

(I-a) $a^{(0)}(0)=b^{(0)}(0)=1$.

(I-b) $a^{(\mu)}(0)=b^{(\mu)}(0)=0$, for all $\mu=1,\dots,n$.

(I-c) For all $\gamma\in\Gamma_M$, $\xi\in\mathbb{R}^d$,

$$\sum_{\mu=0}^{n}\overline{a^{(\mu)}(\xi+\gamma)}b^{(\mu)}(\xi)=\delta_{0,\gamma}.\tag{2.11}$$

Let us express condition (I) in terms of modulation matrices. In (1.37), we already defined modulation matrices from symbol families, provided that $n=m-1$. This notation extends to $n\geq m$ by

$$\mathbf{a}:=\left(a^{(\mu)}(\cdot+\gamma_\nu)\right)_{\substack{\nu=0,\dots,m-1\\ \mu=0,\dots,n}},\qquad \mathbf{b}:=\left(b^{(\mu)}(\cdot+\gamma_\nu)\right)_{\substack{\nu=0,\dots,m-1\\ \mu=0,\dots,n}}.$$

Hence, in contrast to the biorthogonal setting, the modulation matrices are no longer supposed to be square but rectangular. Due to the $\mathbb{Z}^d$-periodicy of trigonometric polynomials, (I-c) is equivalent to

$$\overline{\mathbf{a}}\mathbf{b}^\top = \mathcal{I}_m. \tag{2.12}$$

Since modulation matrices do generally not commute, there is a differences between (2.12) and the condition (1.38) in the biorthogonal context.

The following mixed extension principle was stated in [CHS02, DHRS03]. Due to the results in [Bow00, CSS98], the mild smoothness assumptions on the generators are removed.

Theorem 2.2.1 (MEP). *Let the symbol family* $\{(a^{(\mu)}, b^{(\mu)}) : \mu = 0, \ldots, n\}$ *satisfy* condition (I), *and let* $a^{(0)}$, $b^{(0)}$ *generate refinable functions* $\varphi, \widetilde{\varphi} \in L_2(\mathbb{R}^d)$, *respectively. For* $\mu = 1, \ldots, n$, *define* $\psi^{(\mu)}$, $\widetilde{\psi}^{(\mu)}$ *by* (2.10). *Then*

$$X(\{\psi^{(1)}, \ldots, \psi^{(n)}\}), \quad X(\{\widetilde{\psi}^{(1)}, \ldots, \widetilde{\psi}^{(n)}\})$$

constitute a wavelet bi-frame.

If the assumptions in Theorem 2.2.1 are satisfied and $n = m - 1$, then we obtain a wavelet bi-frame

$$X(\{\psi^{(1)}, \ldots, \psi^{(m-1)}\}), \quad X(\{\widetilde{\psi}^{(1)}, \ldots, \widetilde{\psi}^{(m-1)}\}). \tag{2.13}$$

According to $n = m - 1$, the matrices $\mathbf{a}$ and $\mathbf{b}$ are square, and we have the equivalence

$$\overline{\mathbf{a}}\mathbf{b}^\top = \mathcal{I}_m \quad \text{iff} \quad \mathbf{a}^\top \overline{\mathbf{b}} = \mathcal{I}_m. \tag{2.14}$$

The right-hand side coincides with (1.38). Hence, for $n = m - 1$, (I-c) is equivalent to the condition (1.31) for the construction of biorthogonal wavelets. If the underlying refinable functions φ and $\widetilde{\varphi}$ have biorthogonal integer shifts, then the systems in (2.13) are biorthogonal. By applying Lemma 2.1.6 we verify that they already constitute a pair of biorthogonal wavelet bases. Thus, Theorem 2.2.1 is a direct generalization of Theorem 1.1.11 for the construction of biorthogonal wavelets regarding wavelet bi-frames.

Remark 2.2.2. In biorthogonal constructions, the symbol $b^{(0)}$ has to be dual to $a^{(0)}$. The choice $n \geq m$ in Theorem 2.2.1 has a deep impact since then $b^{(0)}$ is no longer required to be dual to $a^{(0)}$. Moreover, we do not assume any stability of the generators φ and $\widetilde{\varphi}$. Hence, we obtain much more flexibility than in the biorthogonal setting.

2.3 Properties and Optimality Criteria

2.3.1 The Approximation Order of Wavelet Bi-Frames

Given a compactly supported wavelet bi-frame

$$X(\{\psi^{(1)}, \ldots, \psi^{(n)}\}), \quad X(\{\widetilde{\psi}^{(1)}, \ldots, \widetilde{\psi}^{(n)}\})$$

with underlying refinable functions φ and $\widetilde{\varphi}$, respectively, the truncated representation Q_j is a well-defined operator from $L_2(\mathbb{R}^d)$ to V_j, i.e.,

$$Q_j : L_2(\mathbb{R}^d) \to V_j, \quad f \mapsto \sum_{\substack{\mu=1,\ldots,n \\ j'<j, k \in \mathbb{Z}^d}} \left\langle f, \widetilde{\psi}^{(\mu)}_{j',k} \right\rangle \psi^{(\mu)}_{j',k},$$

where $(V_j)_{j\in\mathbb{Z}}$ denotes the multiresolution analysis generated by φ, see (1.51) for the biorthogonal setting. The Definition 1.2.2 about approximation order of a pair of biorthogonal wavelet bases obviously extends to wavelet bi-frames. Since Q_j maps into V_j, the approximation order of the wavelet bi-frame is still bounded by that of the underlying multiresolution analysis. Note that we do not assume that the generator φ is stable since we only use (M-4') on page 34 instead of the stronger condition (M-4) in Definition 1.1.9. Nevertheless, Theorems 1.2.3 and 1.2.4 as well as Corollary 1.2.5 still hold with respect to this weaker assumptions on the generator, cf. [Jia98, Lin05]. Hence, at least for idempotent dilation matrices, the approximation order of the multiresolution analysis is determined by the generator's ability to reproduce polynomials.

In order to obtain a wavelet bi-frame with a high approximation order, we consider two aspects. On the one hand, one requires a high approximation order of the underlying multiresolution analysis, which provides the potential of the wavelet bi-frame. According to Theorem 1.2.7, it can be ensured by the choice of a symbol satisfying sum rules of high order. On the other hand, the bi-frame should provide an *optimal approximation order*, i.e., its approximation order reaches the approximation order of the underlying multiresolution analysis.

Let us address the second aspect. In the sequel, we establish conditions, which ensure that the bi-frame provides optimal approximation order. We say a symbol a has s *vanishing moments* if

$$\partial^\alpha a(0) = 0, \qquad \text{for all } |\alpha| < s.$$

According to (2.10) and the equivalent expression (1.29) in frequency, vanishing moments of the wavelet symbol imply vanishing moments of the associated wavelet. By summarizing the results in [DHRS03], we obtain the following theorem.

Theorem 2.3.1. *Let M be isotropic and let the symbol family*

$$\{(a^{(\mu)}, b^{(\mu)}) : \mu = 0, \ldots, n\}$$

satisfy condition (I) *on page 34, generating a wavelet bi-frame in $L_2(\mathbb{R}^d)$. Let the underlying primal multiresolution analysis exactly provide approximation order s_0. If, for $\mu = 1, \ldots, n$, all products $a^{(\mu)}b^{(\mu)}$ have $2s_1$ vanishing moments, then the approximation order s of the wavelet bi-frame satisfies*

$$\min\{s_0, 2s_1\} \leq s \leq s_0. \tag{2.15}$$

Let the symbol family $\{(a^{(\mu)}, b^{(\mu)}) : \mu = 0, \ldots, n\}$ satisfy condition (I) on page 34. According to the results in [CHR00], if $a^{(0)}$ and $b^{(0)}$ satisfy the sum rules of order s and s', respectively, and provided that $n = m - 1$, then the wavelet symbols have s' and s vanishing moments, respectively. However, the choice $n = m - 1$ provides the restrictions of the biorthogonal setting since it requires the duality (1.26) of $a^{(0)}$ and $b^{(0)}$. Thus, we focus on $n \geq m$, see also Remark 2.2.2. Unfortunately, then the sum rules no longer ensure vanishing moments, and we have to pay extra attention to them. According to Theorems 2.3.1 and 1.2.7, a wavelet bi-frame provides optimal approximation order if the number of vanishing moments is at least half of the sum rule order of the underlying refinable function's symbol.

Next, we address the first aspect, i.e., the approximation order of the underlying multiresolution analysis. According to Corollary 1.2.5 and Theorem 1.2.7, it is determined by

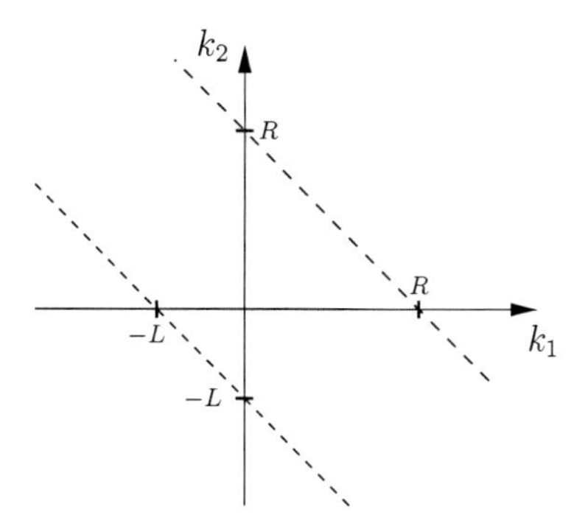

Figure 2.3: The set (2.16) for $d = 2$

the sum rule order of the generator's symbol. In the following, we establish optimality criteria with respect to mask size and sum rule order. Since fundamental refinable functions are on our wish list on page 23, we restrict the following considerations to interpolatory symbols. The next theorem is borrowed from [HJ98].

Theorem 2.3.2. *For $M = 2\mathcal{I}_d$, let an interpolatory symbol satisfy the sum rules of order s. If its mask is supported on*

$$\prod_{i=1}^{d}[-L_i, R_i], \quad \textit{for } L_i, R_i \in \mathbb{N}_0,$$

then

$$s \leq \min_{i=1,\dots,d}\left(\left\lfloor\frac{L_i+1}{2}\right\rfloor + \left\lfloor\frac{R_i+1}{2}\right\rfloor\right).$$

Furthermore, there is a unique univariate interpolatory symbol $\check{c}$, which is supported on $[-2N+1, 2N-1]$ and satisfies the sum rules of order $2N$.

Let us derive similar results for nondyadic dilation. Given nonnegative L, R, we denote

$$[-L, R]_\Sigma := \left\{ (k_1, \dots, k_d)^\top \in \mathbb{Z}^d : -L \leq \sum_{i=1}^{d} k_i \leq R \right\}, \tag{2.16}$$

see Figure 2.3 for a bivariate visualization. Then we obtain the following corollary. It includes the bivariate results in [HJ02], which address the quincunx dilation matrix.

Corollary 2.3.3. *Assume that M satisfies*

$$\Gamma_M = \left\{0, (1/2, \dots, 1/2)^\top\right\}. \tag{2.17}$$

Let c be an interpolatory symbol satisfying the sum rules of order s with respect to M. For $L, R \in \mathbb{N}_0$, let its mask be supported on $[-L, R]_\Sigma$. Then,

$$s \leq \left\lfloor\frac{L+1}{2}\right\rfloor + \left\lfloor\frac{R+1}{2}\right\rfloor$$

holds.

Proof. The univariate symbol

$$\check{c}(\xi_1) := c(\xi_1, \dots, \xi_1)$$

is interpolatory, and its mask is supported on $[-L, R]$. By applying the chain rule of differentiation, we observe that it satisfies the sum rules of order s with respect to dyadic dilation. Theorem 2.3.2 then concludes the proof. □

Theorem 2.3.2 and Corollary 2.3.3 yield optimality criteria for symbols of refinable functions: we say a symbol *provides optimal sum rule order* if its mask support does not allow for a higher order.

2.3.2 Fast Wavelet Frame Transform

The construction of wavelet bi-frames by Theorem 2.2.1 provides the *fast wavelet frame transform*, which is the topic of the present subsection. The core of the transform works identical to the fast wavelet transform in Algorithm 1, but we have to be careful with the input sequence. Given $f \in V_0$ with the expansion

$$f(x) = \sum_{k \in \mathbb{Z}^d} \lambda_k \varphi_{0,k}(x), \tag{2.18}$$

the integer shifts of $\widetilde{\varphi}$ are not necessarily biorthogonal to φ. Hence, we may not conclude

$$\lambda_k = \langle f, \widetilde{\varphi}_{0,k} \rangle, \quad k \in \mathbb{Z}^d.$$

In order to derive the inner product from the coefficients in (2.18), we introduce the *bracket product* of $\widehat{\varphi}$ and $\widehat{\widetilde{\varphi}}$, i.e.,

$$\left[\widehat{\varphi}, \widehat{\widetilde{\varphi}}\right] := \sum_{k \in \mathbb{Z}^d} \widehat{\varphi}(\cdot - k)\overline{\widehat{\widetilde{\varphi}}(\cdot - k)}. \tag{2.19}$$

Since both φ and $\widetilde{\varphi}$ are compactly supported, Poisson's summation formula yields that (2.19) is a trigonometric polynomial, cf. [JM90]. According to [DHRS03], we obtain

$$\left(\langle f, \widetilde{\varphi}_{0,k} \rangle\right)_{k \in \mathbb{Z}^d} = \left[\widehat{\varphi}, \widehat{\widetilde{\varphi}}\right] * (\lambda_k)_{k \in \mathbb{Z}^d}, \tag{2.20}$$

where we apply the filter bank notation in Appendix A.23, i.e., one convolves the Fourier coefficients of the trigonometric polynomial with the sequence.

For the fast wavelet frame transform, the input sequence $H_0^{(0)}$ is expected either to be the coefficient sequence in (2.18) or directly the inner products

$$H_0^{(0)}(k) = \langle f, \widetilde{\varphi}_{0,k} \rangle, \quad k \in \mathbb{Z}^d. \tag{2.21}$$

Then, for $j < 0$, $\mu = 1, \dots, n$, the transform computes

$$H_j^{(0)}(k) = \langle f, \widetilde{\varphi}_{j,k} \rangle,$$
$$H_j^{(\mu)}(k) = \left\langle f, \widetilde{\psi}_{j,k}^{(\mu)} \right\rangle,$$

see Algorithm 2. Note that in case of $H_0^{(0)} = (\lambda_k)_{k\in\mathbb{Z}^d}$ as the input sequence of the decomposition, we convolve it with $\left[\widehat{\varphi}, \widehat{\widetilde{\varphi}}\right]$. Then the reconstruction requires the deconvolution of $\left[\widehat{\varphi}, \widehat{\widetilde{\varphi}}\right]$. From a computational point of view, the deconvolution is unproblematic since it can be accomplished by some predefined deconvolution method in matlab. Alternatively, one may apply the fast Fourier transform. Then deconvolution means division on the Fourier domain. The computation of $\left[\widehat{\varphi}, \widehat{\widetilde{\varphi}}\right]$ is more expensive, but it is only required once.

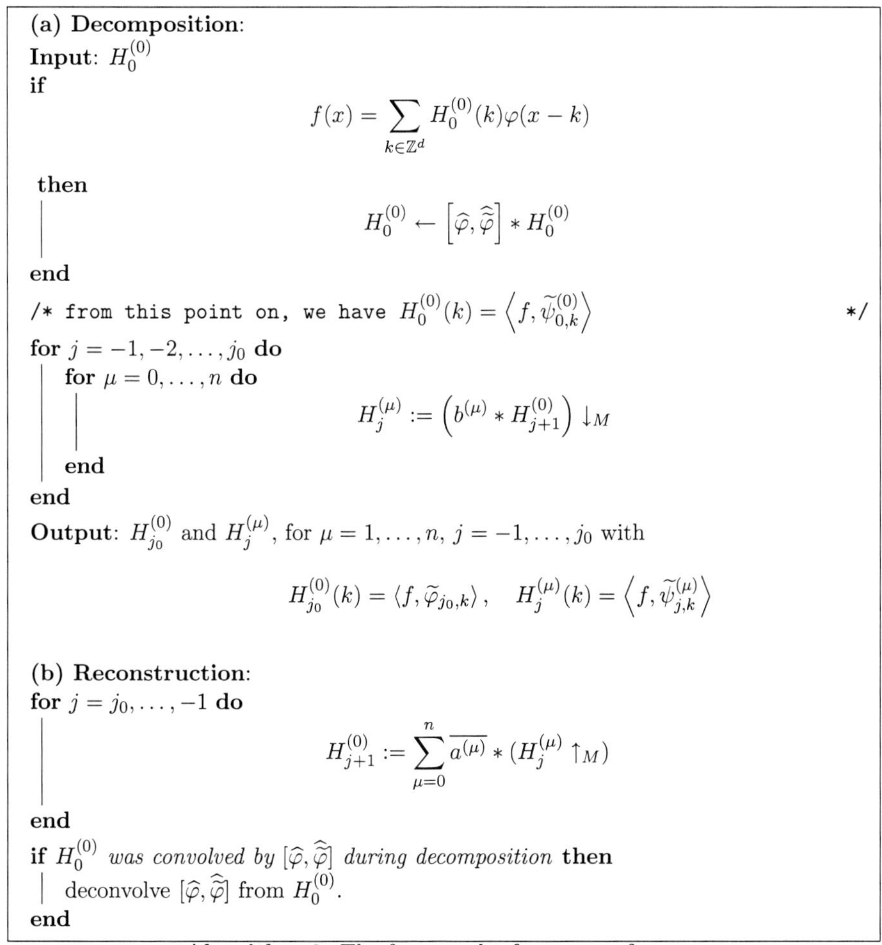

(a) Decomposition:
Input: $H_0^{(0)}$
if

$$f(x) = \sum_{k\in\mathbb{Z}^d} H_0^{(0)}(k)\varphi(x-k)$$

then

$$H_0^{(0)} \leftarrow \left[\widehat{\varphi}, \widehat{\widetilde{\varphi}}\right] * H_0^{(0)}$$

end
`/* from this point on, we have` $H_0^{(0)}(k) = \left\langle f, \widetilde{\psi}_{0,k}^{(0)} \right\rangle$ `*/`
for $j = -1, -2, \ldots, j_0$ **do**
  for $\mu = 0, \ldots, n$ **do**

$$H_j^{(\mu)} := \left(b^{(\mu)} * H_{j+1}^{(0)}\right) \downarrow_M$$

  end
end
Output: $H_{j_0}^{(0)}$ and $H_j^{(\mu)}$, for $\mu = 1, \ldots, n$, $j = -1, \ldots, j_0$ with

$$H_{j_0}^{(0)}(k) = \langle f, \widetilde{\varphi}_{j_0,k} \rangle, \quad H_j^{(\mu)}(k) = \left\langle f, \widetilde{\psi}_{j,k}^{(\mu)} \right\rangle$$

(b) Reconstruction:
for $j = j_0, \ldots, -1$ **do**

$$H_{j+1}^{(0)} := \sum_{\mu=0}^{n} \overline{a^{(\mu)}} * (H_j^{(\mu)} \uparrow_M)$$

end
if $H_0^{(0)}$ *was convolved by* $[\widehat{\varphi}, \widehat{\widetilde{\varphi}}]$ *during decomposition* **then**
  deconvolve $[\widehat{\varphi}, \widehat{\widetilde{\varphi}}]$ from $H_0^{(0)}$.
end

Algorithm 2: The fast wavelet frame transform

If we do not apply any convolution and deconvolution of $\left[\widehat{\varphi}, \widehat{\widetilde{\varphi}}\right]$, then we still have a filter bank with exact reconstruction, see Figure 2.4. However, we no longer have any interpretation of the sequences $H_j^{(\mu)}$ as inner products. Note that Figure 1.4 and Figure 2.4 only differ by the number of channels.

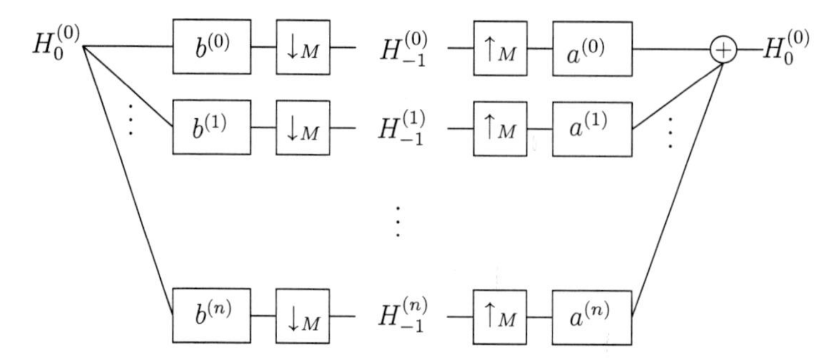

Figure 2.4: The filter bank scheme of the fast wavelet frame transform

In order to bound the complexity of the transform, one must ask for the minimal number of channels. According to condition (I) on page 34, the number of channels is at least m, i.e., $n = m - 1$. However, then the symbol $b^{(0)}$ has to be dual to $a^{(0)}$, and the wavelet bi-frame suffers from the restrictions of the biorthogonal setting, see Chapter 1.

Let us address the desirable case $m = 2$ in more detail. According to our wish list on page 23, let the underlying primal refinable function be fundamental, and let the dual refinable function coincide with the primal one. Then an application of Lemma 1.3.1 yields $n \geq 2$:

Lemma 2.3.4. *Let the symbol family*

$$\left\{\left(a^{(0)}, a^{(0)}\right), \left(a^{(1)}, b^{(1)}\right), \ldots, \left(a^{(n)}, b^{(n)}\right)\right\}$$

satisfy condition (I) *on page 34. If $a^{(0)}$ generates a fundamental refinable function, then $n \geq 2$.*

Proof. Since $n \geq m - 1$, we merely consider $m = 2$, and we assume $n = 1$. Then the associated polyphase matrices are square, and (2.14) yields

$$\mathbf{a}^\top \overline{\mathbf{b}} = \mathcal{I}_m.$$

Thus, Lemma 1.3.1 with $a^{(0)} = b^{(0)}$ and $\varphi = \widetilde{\varphi}$ imply that the underlying refinable function is not fundamental. □

Lemma 2.3.4 provides the following optimality condition. For $m = 2$, let a wavelet bi-frame be constructed by Theorem 2.2.1, and suppose that primal and dual wavelets are obtained from one single fundamental refinable function. We say the wavelet bi-frame *provides an optimal number of wavelets* if it is generated by only two wavelets.

2.3.3 Symmetry

We already mentioned in Section 1.2.2, that symmetric wavelets are desired in image and signal analysis. With regard to the construction of wavelet bi-frames by Theorem 2.2.1, it is important to express symmetry in terms of the associated symbols. Following Han in [Han04], we consider symmetry with respect to a so-called *symmetry group* $\mathcal{G}$, i.e., $\mathcal{G}$ is a finite subgroup of

$$\left\{U \in \mathbb{Z}^{d\times d} : |\det(U)| = 1\right\}$$

such that

$$MUM^{-1} \in \mathcal{G}, \quad \text{for all } U \in \mathcal{G}. \tag{2.22}$$

A symbol a is called *$\mathcal{G}$-symmetric* if there exists $p \in \mathbb{R}^d$ such that

$$a_{U(k-p)+p} = a_k, \quad \text{for all } U \in \mathcal{G},\, k \in \mathbb{Z}^d.$$

Thus, we implicitly impose $(\mathcal{I}_d - U)p \in \mathbb{Z}^d$, for all $U \in \mathcal{G}$. A function f is called *$\mathcal{G}$-symmetric* if there is $q \in \mathbb{R}^d$ such that

$$f(U(x-q)+q) = f(x), \quad \text{for all } U \in \mathcal{G},\, x \in \mathbb{R}^d.$$

The role of the symmetry group can be described as follows: the elements of $\mathcal{G}$ express the symmetry, and the additional requirement (2.22) transfers symbol symmetries into symmetries of the refinable function and the wavelets. According to the results in [Han04], we have the following theorem:

Theorem 2.3.5. *Given a symbol a with $a(0) = 1$, let φ be its generated refinable function. If a is $\mathcal{G}$-symmetric, then φ is also $\mathcal{G}$-symmetric. Moreover, if a and the associated wavelet symbol are $\mathcal{G}$-symmetric, then the wavelet is also $\mathcal{G}$-symmetric.*

Theorem 2.3.5 yields some good news for the construction of wavelets by Theorem 2.2.1. Symmetric wavelets can be obtained by the choice of symmetric symbols. We say a wavelet bi-frame *provides optimal symmetry* if the wavelets inherit the full symmetries of the underlying refinable function in the sense of Theorem 2.3.5.

2.3.4 Smoothness

For refinable functions, the importance of smoothness is twofold. First, smoothness can be used as a sufficient condition for approximation order of the multiresolution analysis since it yields reproduction of polynomials, which provides a high approximation order, cf. Theorem 1.2.6 and Corollary 1.2.5.

Second, certain smoothness is necessary for the characterization of smoothness spaces since wavelets are required to be contained in the space. In Subsection 1.2.3, we already mentioned that pairs of biorthogonal wavelet bases characterize certain function spaces, i.e., they constitute bases for the space und a weighted sequence norm of wavelet coefficients yields an equivalent norm. Contrary to Chapter 1, the present chapter addresses the weaker concept of wavelet bi-frames. Recently, Borup, Gribonval, and Nielsen extended the characterization of function spaces from biorthogonal wavelets to dyadic wavelet bi-frames, see [BGN04] for details. Moreover, in Chapter 5 of the present thesis, we generalize their results regarding wavelet bi-frames with isotropic dilation matrices. The main requirements for such a characterization are smoothness and vanishing moments.

'Smoothness' and support size are competing properties. In this section, we derive optimality criteria in the form:

"Given a support size, then the refinable function can reach at most a certain smoothness."

Following Han and Jia in [Han99, HJ02], we measure smoothness in the scale of Lipschitz spaces $\mathrm{Lip}\left(s, L_p(\mathbb{R}^d)\right)$, $1 \leq p \leq \infty$, cf. Appendix A.1. For $f \in L_p(\mathbb{R}^d)$, we consider the *L_p-critical exponent*

$$s_p(f) := \sup\left\{s \geq 0 : f \in \mathrm{Lip}\left(s, L_p(\mathbb{R}^d)\right)\right\}. \tag{2.23}$$

This means, for $p = 2$, we essentially measure Sobolev regularity and, for $p = \infty$, one estimates Hölder smoothness, i.e.,

$$s_2(f) = \sup\left\{s \geq 0 : f \in W^s(L_2(\mathbb{R}^d))\right\}, \tag{2.24}$$

$$s_\infty(f) = \sup\left\{s \geq 0 : f \in \mathcal{C}^s(\mathbb{R}^d)\right\}, \tag{2.25}$$

see Appendix A.1 for details. The first result for dyadic dilation is borrowed from [Han99].

Theorem 2.3.6. *For $M = 2I_d$, let an interpolatory symbol a generate the fundamental refinable function φ. If a is supported on $[-2N+1, 2N-1]^d$ and satisfies the sum rules of order $2N$, then*

$$s_p(\varphi) \leq s_p(\varphi_0), \quad 1 \leq p \leq \infty, \tag{2.26}$$

where φ_0 is the compactly supported refinable function of the univariate dyadic interpolatory symbol $\check{c}$, whose mask is supported on $[-2N+1, 2N-1]$, satisfying the sum rules of order $2N$.

According to Theorem 2.3.2, the symbol $\check{c}$ is uniquely determined. Then φ_0 in (2.26) is unique up to multiplication with a constant, cf. [CDM91]. For $N = 2$, we have $s_\infty(\varphi_0) = 2$, see [Han99]. This yields the following corollary, which is also contained in [Han99]:

Corollary 2.3.7. *Let an interpolatory symbol a generate a fundamental refinable function φ with $M = 2I_d$. If a is supported on $[-3, 3]^d$, then*

$$s_\infty(\varphi) \leq 2.$$

Theorem 2.3.6 and Corollary 2.3.7 address dyadic dilation. The relation of smoothness and support size is much more complicated for nondyadic dilation. In the sequel, we derive optimality criteria for specific dilation matrices mentioned in [Han02]. These matrices can be defined in arbitrary dimensions, and they satisfy $m = 2$. Hence, they allow for a minimal number of wavelets. Then in Chapters 3 and 4, we construct wavelet bi-frames with their scalings: for $d = 2, 3$, let

$$M = \begin{pmatrix} -1 & 1 \\ 1 & 1 \end{pmatrix}, \qquad M = \begin{pmatrix} 0 & 2 & 1 \\ -1 & -1 & 0 \\ 1 & 1 & 1 \end{pmatrix}. \tag{2.27}$$

For $d > 3$, we define

$$M = \begin{pmatrix} 0 & 2 & 1 & \cdots\cdots & & 1 \\ \vdots & \ddots & 1 & 0 & \cdots & 0 \\ \vdots & & \ddots & \ddots & \ddots & \vdots \\ 0 & \cdots\cdots & & 0 & 1 & 0 \\ -1 & \cdots\cdots & & & -1 & 0 \\ 1 & \cdots\cdots & & & & 1 \end{pmatrix}. \tag{2.28}$$

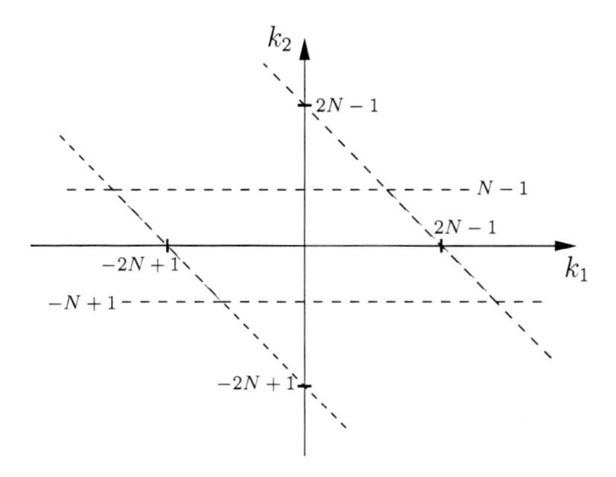

Figure 2.5: The support set in (2.30) for $d = 2$

Then $M^d = 2\mathcal{I}_d$, and M generates the *checkerboard lattice*, i.e.,

$$M\mathbb{Z}^d = \Big\{(k_1, \ldots, k_d)^\top \in \mathbb{Z}^d : \sum_{i=1}^{d} k_i \in 2\mathbb{Z}\Big\}. \tag{2.29}$$

Note that the bivariate Checkerboard lattice is also generated by the box spline and quincunx dilation matrices M_b and M_q, respectively, cf. (1.6).

Since Γ_M can be chosen as in (2.17), the sum rule estimates of Corollary 2.3.3 are applicable. Moreover, under slightly stronger assumptions, we also obtain estimates with respect to smoothness.

Proposition 2.3.8. *Given M by (2.27), (2.28), let a symbol a satisfy the sum rules of order $2N$, and let it generate a fundamental refinable function φ. If its mask is supported on*

$$[-2N+1, 2N-1]_\Sigma \cap \Big\{k \in \mathbb{Z}^d : \Big|\sum_{i=l}^{d} k_i\Big| \leq N-1, \;\; l = 2, \ldots, d\Big\}, \tag{2.30}$$

then

$$s_p(\varphi) \leq s_p(\varphi_0), \quad 1 \leq p \leq \infty, \tag{2.31}$$

with φ_0 as in Theorem 2.3.6.

For $d = 2$, Proposition 2.3.8 is already stated in [HJ02]. The concrete choice of M allows the extension to higher dimensions, see Figure 2.5 for a visualization of the support set (2.30) with $d = 2$.

Proof. Due to $M^d = 2\mathcal{I}_d$, φ is also refinable with respect to the symbol

$$c(\xi) := a(\xi) \cdot a(M^\top \xi) \cdots a(M^{\top^{d-1}} \xi)$$

with dyadic dilation. Define the univariate symbol $\check{c}$ by

$$\check{c}(\xi_1) := c(\xi_1, \ldots, \xi_1).$$

We only verify that $\check{c}$ is the unique interpolatory symbol in Theorem 2.3.2. Then following the lines in [HJ02] yields Proposition 2.3.8.

First, we verify

$$a(0, \xi_1, \ldots, \xi_1) = 1, \ \ldots, \ a(0, \ldots, 0, \xi_1) = 1, \quad \text{for all } \xi_1 \in \mathbb{R}. \tag{2.32}$$

Due to [HJ02], (2.32) holds provided that $d = 2$. For arbitrary d, we reduce to the bivariate setting by

$$\widetilde{a}(\xi_1, \xi_2) := a(\xi_1, \ldots, \xi_1, \xi_2, \ldots, \xi_2),$$

where we can vary the numbers of appearing ξ_1, ξ_2. Both should appear at least once. The symbol $\widetilde{a}$ is interpolatory, and its mask is supported on (2.30) with $d = 2$. By applying the chain rule of differentiation, it satisfies the sum rules of order $2N$. This concludes the proof of (2.32).

According to (2.27), (2.28), and (2.32), we have $\check{c}(\xi_1) = a(\xi_1, \ldots, \xi_1)$. Thus, it is interpolatory, and it satisfies the sum rules of order $2N$. By applying (2.30), the mask of $\check{c}$ is supported on $[-2N + 1, 2N - 1]$. Hence, it is the unique symbol in Theorem 2.3.2, which concludes the proof. □

We derived optimality criteria with respect to smoothness and support size: let us say a refinable function φ has *optimal smoothness* if either Theorem 2.3.6 or Proposition 2.3.8 is applicable and the associated inequality (2.26) or (2.31), respectively, becomes an equality.

2.3.5 Summary of Optimality Criteria

In the previous subsections, we established optimality criteria for wavelet bi-frames, see Table 2.2 for a summary of their references. On the one hand, we purely consider optimality in terms of the underlying refinable function:

Mask Size vs. Sum Sule Order / Smoothness

Mask size and sum rule order are competing properties. Then Theorem 2.3.2 and Corollary 2.3.3 provide statements about maximal sum rule order with respect to some given mask size. Generally, mask size and smoothness are also competing properties. In order to obtain smoothness bounds for a given mask size, a univariate refinable function plays the role of a reference function, whose smoothness is a benchmark. Then Theorem 2.3.6, Corollary 2.3.7, and Proposition 2.3.8 provide statements about maximal Lipschitz smoothness with respect to some given mask size.

On the other hand, we examine whether the wavelets exploit the potential of the underlying refinable function:

Approximation Order / Symmetry / Number of Wavelets

The refinable function generates a multiresolution analysis which provides a certain approximation order. We require a wavelet bi-frame that reaches this approximation order. Theorem 2.3.1 tells us that the bi-frame provides optimal approximation order if the number of vanishing moments is at least the half of the order of sum rules. This provides a main benchmark for our constructions in Chapters 3 and 4.

property	Reference
sum rules	Theorem 2.3.2, Corollary 2.3.3
smoothness	Theorem 2.3.6, Corollary 2.3.7, Proposition 2.3.8
card(wavelets)	Lemma 2.3.4
approx. order	Theorem 2.3.1
symmetry	Theorem 2.3.5

Table 2.2: Optimality criteria

Given an underlying refinable function with certain symmetries, then the wavelets should preserve these symmetries. According to Theorem 2.3.5, this can be ensured by a careful choice of the symbols.

In order to bound the complexity of the fast wavelet frame transform, we desire a minimal number of wavelets. Provided that all wavelets are derived from one single fundamental refinable function, one requires at least two wavelets, see Lemma 2.3.4.

Chapter 3

Compactly Supported Wavelet Bi-Frames Obtained by Convolution

The present chapter is dedicated to the construction of wavelet bi-frames. Our aim is the verification that the more flexible bi-frame approach can overcome the restrictions of wavelet bases, and we look for wavelet bi-frames providing all items on our wish list in Section 1.3.

First, we establish the general construction procedure based on the mixed extension principle. Roughly speaking, our approach is based on the convolution of a pair of biorthogonal wavelets bases. Since we do not require any smoothness, their choice can satisfy all other items of our wish list. By some convolution process, we derive a wavelet bi-frame. It essentially preserves the properties of the biorthogonal wavelets, and smoothness is added by the convolution. In fact, the procedure works under weaker assumptions: we do not require a pair of biorthogonal wavelet bases, but only symbols satisfying the conditions (1.31). The underlying refinable functions φ and $\widetilde{\varphi}$ do not even need to be contained in $L_2(\mathbb{R}^d)$. We merely suppose that their convolution is in $L_2(\mathbb{R}^d)$. Since we consider compactly supported functions, membership in $L_2(\mathbb{R}^d)$ means some kind of smoothness condition. Hence, these weaker assumptions allow for extremely small supports. It should be mentioned that the convolved refinable function $\varphi * \widetilde{\varphi}$ can be very smooth, although φ or $\widetilde{\varphi}$ are not even contained in $L_2(\mathbb{R}^d)$. Primal and dual wavelets of the resulting bi-frame are obtained from this single refinable function $\varphi * \widetilde{\varphi}$. Hence, they are contained in the same smoothness class.

Then, we discuss the choice of the symbols satisfying the conditions (1.31). Finally, our approach provides optimality with respect to the criteria summarized in Subsection 2.3.5. The underlying refinable function can be chosen to be fundamental with an optimal sum rule order. The general procedure provides also an optimal approximation order, i.e., the wavelets have sufficient vanishing moments. As a highlight, we construct real-valued wavelet bi-frames in any dimensions with arbitrary smoothness and an arbitrary number of vanishing moments. The bi-frame is generated by only three wavelets and they are symmetric about a point. Moreover, the underlying refinable function provides optimal smoothness. We also derive wavelet bi-frames with respect to the quincunx dilation matrix, and we construct a wavelet bi-frame from a fundamental version of a bivariate box spline. Our bi-frames possess similar or even better properties with significantly smaller mask sizes than comparative biorthogonal wavelet bases from [CHR00, HJ02, HR02].

3.1 A Construction by the Mixed Extension Principle

First, we verify that condition (I) on page 34 is invariant under a certain multiplication process:

Theorem 3.1.1. *Let the symbol family* $\{(a^{(\mu)}, b^{(\mu)}) : \mu = 0, \ldots, n_1\}$ *satisfy* condition (I), *and let another family* $\{(c^{(\nu)}, d^{(\nu)}) : \nu = 0, \ldots, n_2\}$ *satisfy* (I-a), (I-b), *and at least the equation in* (I-c) *for* $\gamma = 0$.

(a) *Then the family*

$$\left\{\left(a^{(\mu)}c^{(\nu)}, b^{(\mu)}d^{(\nu)}\right) : \mu = 0, \ldots, n_1, \ \ \nu = 0, \ldots, n_2\right\} \tag{3.1}$$

satisfies condition (I).

(b) *For* $\nu = 0, \ldots, n_2$ *and* $\mu = 1, \ldots, n_1$, *define*

$$\begin{array}{rclrcl} \breve{a}^{(\nu)}(\xi) & := & a^{(0)}(\xi)c^{(\nu)}(M^\top\xi), & \breve{b}^{(\nu)}(\xi) & := & b^{(0)}(\xi)d^{(\nu)}(M^\top\xi), \\ \breve{a}^{n_2+\mu} & := & a^{(\mu)}, & \breve{b}^{n_2+\mu} & := & b^{(\mu)}. \end{array} \tag{3.2}$$

Then also the family

$$\left\{\left(\breve{a}^{(\mu)}, \breve{b}^{(\mu)}\right) : \mu = 0, \ldots, n_1 + n_2\right\}$$

satisfies condition (I).

Theorem 3.1.1 extends some known results in the wavelet frame literature: for the choice $n_1 = n_2$ and

$$a^{(\mu)} = b^{(\mu)} = c^{(\mu)} = d^{(\mu)}, \quad \mu = 0, \ldots, n_1,$$

part (a) becomes Lemma 3.2 in [Han03a]. There, it is applied to the construction of tight wavelet frames. For $a^{(\mu)} = b^{(\mu)}$, and $c^{(\nu)} = d^{(\nu)}$, part (b) reduces to the inductive construction algorithm given in [GR98, RS98]. These restrictions have some crucial drawbacks: in order to obtain a minimal number of wavelets, one chooses $n_1 = m - 1$. Then $a^{(\mu)} = b^{(\mu)}$, for $\mu = 0, \ldots, m-1$, means that the symbol family has to satisfy the conditions for an orthogonal wavelet bases. This causes the same problems as already mentioned in Section 1.3: for many dilation matrices, it means a lack of smoothness and symmetry of the underlying functions.

Finally, the benefit of the more flexible Theorem 3.1.1 is the following: a choice $a^{(\mu)} \neq b^{(\mu)}$ allows for smooth and symmetric wavelet bi-frames. In order to clarify the last statement, we formulate Theorem 3.1.1 with respect to two identical symbol families of length $n = m - 1$, and then we translate the symbol families into refinable functions and wavelets.

Corollary 3.1.2. *Let the symbol family* $\{(a^{(\mu)}, b^{(\mu)}) : \mu = 0, \ldots, m-1\}$ *satisfy* condition (I) *on page 34. Then also the family*

$$\left\{\left(a^{(\mu)}b^{(\nu)}, b^{(\mu)}a^{(\nu)}\right) : \mu, \nu = 0, \ldots, m-1\right\} \tag{3.3}$$

satisfies condition (I).

Under the assumptions of Corollary 3.1.2, let φ and $\widetilde{\varphi}$ be generated by $a^{(0)}$ and $b^{(0)}$, respectively. Then, due to the Fourier transformed refinement equation (1.23), their convolution $\varphi * \widetilde{\varphi}$ is refinable with respect to the product $a^{(0)}b^{(0)}$. We denote the wavelets corresponding to $a^{(\mu)}$ and $b^{(\mu)}$ as usual by $\psi^{(\mu)}$ and $\widetilde{\psi}^{(\mu)}$, $\mu = 1, \dots, n$, respectively. Let us also denote

$$\psi^{(0)} := \varphi, \qquad \widetilde{\psi}^{(0)} := \widetilde{\varphi}.$$

Then the wavelets according to $a^{(\mu)}b^{(\nu)}$ are given by $\psi^{(\mu)} * \widetilde{\psi}^{(\nu)}$. If the convolution $\varphi * \widetilde{\varphi}$ is additionally contained in $L_2(\mathbb{R}^d)$, then

$$X(\{\psi^{(\mu)} * \widetilde{\psi}^{(\nu)} : \mu, \nu = 0, \dots, n, (\mu, \nu) \neq (0,0)\}) \tag{3.4}$$

constitutes a wavelet frame in $L_2(\mathbb{R}^d)$, cf. Theorem 2.2.1. Its underlying refinable function is $\varphi * \widetilde{\varphi}$, and the reconstruction formula

$$f = \sum_{\substack{\mu,\nu=0,\dots,n \\ (\mu,\nu)\neq(0,0) \\ j\in\mathbb{Z}, k\in\mathbb{Z}^d}} \left\langle f, (\psi^{(\mu)} * \widetilde{\psi}^{(\nu)})_{j,k} \right\rangle (\psi^{(\nu)} * \widetilde{\psi}^{(\mu)})_{j,k}$$

holds for all $f \in L_2(\mathbb{R}^d)$. Hence, a permutation of (3.4) provides a dual wavelet frame.

The number of resulting wavelets equals $m^2 - 1$. In order to minimize their number, one chooses a dilation matrix with $m = 2$, and one obtains a wavelet bi-frame with 3 wavelets.

Remark 3.1.3. Corollary 3.1.2 may provide a great tool for the construction of wavelet bi-frames: the start symbols $\{(a^{(\mu)}, b^{(\mu)}) : \mu = 0, \dots, m-1\}$ can be chosen with extremely small supports and much symmetry if we abandon smoothness of their underlying functions. Then the multiplication of symbols is equivalent to the convolution of wavelets and refinable functions. By a careful choice of the start symbols, this convolution can yield very smooth functions and symmetry can be preserved. Finally, this construction has the potential for providing all items on our wish list since convolution does not enlarge the supports too much.

We still have to present the proof of Theorem 3.1.1:

Proof of Theorem 3.1.1. The conditions (I-a) and (I-b) are obviously satisfied.

(a): Let $\gamma \in \Gamma_M$. By applying (I-c) for both given families, we obtain for the family (3.1)

$$\begin{aligned}\sum_{\mu,\nu} \overline{a^{(\mu)}(\xi+\gamma)c^{(\nu)}(\xi+\gamma)}b^{(\mu)}(\xi)d^{(\nu)}(\xi) &= \sum_{\mu,\nu} \overline{a^{(\mu)}(\xi+\gamma)}b^{(\mu)}(\xi)\overline{c^{(\nu)}(\xi+\gamma)}d^{(\nu)}(\xi) \\ &= \delta_{0,\gamma} \sum_{\nu} \overline{c^{(\nu)}(\xi+\gamma)}d^{(\nu)}(\xi) = \delta_{0,\gamma}.\end{aligned}$$

(b): According to $M^\top \gamma \in \mathbb{Z}^d$, the definitions (3.2) yield

$$\begin{aligned}
\sum_{\mu=0}^{n_1+n_2} \overline{\breve{a}^{(\mu)}(\xi+\gamma)\breve{b}^{(\mu)}(\xi)} &= \sum_{\nu=0}^{n_2} \overline{a^{(0)}(\xi+\gamma)c^{(\nu)}(M^\top\xi)b^{(0)}(\xi)d^{(\nu)}(M^\top\xi)} \\
&\quad + \sum_{\mu=1}^{n_1} \overline{a^{(\mu)}(\xi+\gamma)b^{(\mu)}(\xi)} \\
&= \overline{a^{(0)}(\xi+\gamma)b^{(0)}(\xi)} \sum_{\nu=0}^{n_2} \overline{c^{(\nu)}(M^\top\xi)d^{(\nu)}(M^\top\xi)} \\
&\quad + \sum_{\mu=1}^{n_1} \overline{a^{(\mu)}(\xi+\gamma)b^{(\mu)}(\xi)} \\
&= \overline{a^{(0)}(\xi+\gamma)b^{(0)}(\xi)} + \sum_{\mu=1}^{n_1} \overline{a^{(\mu)}(\xi+\gamma)b^{(\mu)}(\xi)} \\
&= \sum_{\mu=0}^{n_1} \overline{a^{(\mu)}(\xi+\gamma)b^{(\mu)}(\xi)} = \delta_{0,\gamma}. \qquad \square
\end{aligned}$$

3.2 Finding Start Symbols

In this section, we address the problem of finding appropriate start symbols for Corollary 3.1.2, i.e., we are looking for a symbol family $\{(a^{(\mu)}, b^{(\mu)}) : \mu = 0, \dots, m-1\}$ satisfying condition (I) on page 34. Due to (2.14), this requires that $b^{(0)}$ is dual to $a^{(0)}$.

3.2.1 Wavelet Symbols

Let us suppose that we have already chosen some symbol $a^{(0)}$ and a dual symbol $b^{(0)}$ (we address their choice in Subsection 3.2.2). In order to find additional symbols satisfying (2.14), we need to solve the matrix extension problem as discussed in Section 1.1.4. Let us recall the main facts for the solution of the problem. In Section 1.1.4, we described that the Theorem of Quillen and Suslin ensures the existence of a solution. However, determining a specific solution generally requires some computer algebra software as well as large computational ressources. Moreover, the algorithms do not ensure minimal support sizes of their solutions, see for instance [Par95]. In order to circumvent such difficulties, one makes some additional assumptions on the symbol of the refinable function: provided that $a^{(0)}$ is interpolatory, see (1.48), Lemma 1.1.20 yields a symbol family satisfying condition (I) on Page 34. However, the restriction to interpolatory $a^{(0)}$ is quite strong. Fortunately, at least for $m = 2$, the limitation can be discarded since Example 1.1.19 also provides a solution $a^{(1)}$ and $b^{(1)}$ for noninterpolatory $a^{(0)}$. Hence, the desirable setting $m = 2$ allows for more flexibility than $m > 2$.

In order to reach an optimal approximation order, we need a wavelet bi-frame with sufficient vanishing moments. This can be ensured by the sum rule order of $a^{(0)}$ and $b^{(0)}$: if both satisfy the sum rules of order s, then a result in [CHR00] yields that all start wavelet symbols $a^{(\mu)}$ and $b^{(\mu)}$, $\mu = 1, \dots, m-1$, have s vanishing moments. Since Corollary 3.1.2 preserves this number and the product $a^{(0)}b^{(0)}$ satisfies the sum rules of order $2s$, the wavelets' number of vanishing moments is at least half of the refinable function's reproduction of polynomials. Thus, according to Theorem 2.3.1, the resulting wavelet bi-frame can provide an optimal approximation order.

3.2.2 The Dual Symbol

In this subsection, we address the choice of $a^{(0)}$ and $b^{(0)}$. If $a^{(0)}$ satisfies the sum rules of order s, then, according to the ideas about vanishing moments in Subsection 3.2.1, we are interested in $b^{(0)}$ also satisfying the sum rules of order s. Since we do not require that $b^{(0)}$ generates a smooth refinable function, it can be chosen with very small mask size. In fact, this is our great advantage compared to biorthogonal constructions, where one needs at least that its refinable function is contained in $L_2(\mathbb{R}^d)$. We merely require that the product $a^{(0)}b^{(0)}$ generates a smooth refinable function.

Next, we address the structure of dual symbols. Given a symbol a and an initial dual symbol u, one easily verifies that, for all symbols c,

$$b(\xi) = u(\xi) + c(\xi) - u(\xi) \sum_{\gamma \in \Gamma_M} (c\overline{a})(\xi + \gamma) \tag{3.5}$$

is also dual to a. If a is interpolatory, then one can choose $u \equiv 1$, and (3.5) reduces to

$$b(\xi) = 1 + c(\xi) - \sum_{\gamma \in \Gamma_M} (c\overline{a})(\xi + \gamma). \tag{3.6}$$

The construction of b in (3.6) follows the spirit of the so-called lifting scheme as discussed in a sequence of papers [Swe96, Swe97, DS98, KS00]. The parallel to the philosophy of the lifting concept may be described as follows: from the trivial dual symbol $u \equiv 1$, we construct another dual symbol b with hopefully better properties. In other words, we lift $u \equiv 1$, and we obtain a better symbol b.

It turns out that (3.6) essentially provides a representation of all dual symbols of a:

Theorem 3.2.1. *Let an interpolatory symbol a satisfy the sum rules of order s. Then the following holds:*

(a) *If an interpolatory symbol c satisfies the sum rules of order s, then*

$$b(\xi) = c(\xi) + 1 - \sum_{\gamma \in \Gamma_M} (c\overline{a})(\xi + \gamma) \tag{3.7}$$

is dual to a, and it satisfies the sum rules of order s.

(b) *Let some symbol b be dual to a, and let it satisfy the sum rules of order s. Then*

$$c(\xi) = b(\xi) + \frac{1}{m}\Big(1 - \sum_{\gamma \in \Gamma_M} b(\xi + \gamma)\Big) \tag{3.8}$$

is interpolatory, it satisfies the sum rules of order s, and b can be represented by (3.7).

A natural choice in Theorem 3.2.1 is $c = a$ or $c = \overline{a}$. Then b generally does not generate a refinable function in $L_2(\mathbb{R}^d)$. Fortunately, the product ab often yields a smooth refinable function as required for our construction.

Remark 3.2.2. The CBC algorithm proposed in [Han99, HR02] for the construction of dual symbols also addresses noninterpolatory symbols as well as higher orders of sum rules for the dual symbol. Thus, it is much more general. However, Theorem 3.2.1 can be obtained independently, and it is formulated and proven strictly in terms of symbols instead of masks. We think the representation above provides some insight in the strucure of the dual symbol as we shall explain in the following. Recall that a symbol is called *orthogonal* if its conjugate is dual, i.e.,

$$\sum_{\gamma\in\Gamma_M} |a(\xi+\gamma)|^2 = 1.$$

For $c = a$, (3.7) reads as

$$b(\xi) = a(\xi) + 1 - \sum_{\gamma\in\Gamma_M} |a(\xi+\gamma)|^2.$$

Thus, b equals a up to the amount which a lacks in order to be orthogonal.

Proof of Theorem 3.2.1. (a) As already mentioned, it can directly be verified that b is indeed dual to a. For interpolatory a and c, the sum rules of order s imply

$$\partial^\alpha a(0) = \delta_{0,\alpha} \quad \text{and} \quad \partial^\alpha c(0) = \delta_{0,\alpha}, \quad \text{for all } |\alpha| < s. \tag{3.9}$$

Furthermore, (3.9) and the sum rules of a and c imply the sum rules of order s for b.

(b) A short calculation yields that c is interpolatory. Let us verify the sum rules for c. By applying the duality (1.26), we obtain

$$\partial^\alpha \sum_{\gamma\in\Gamma_M} (b\overline{a})(\cdot+\gamma)\big|_0 = \delta_{0,\alpha}. \tag{3.10}$$

Since a and b satisfy the sum rules of order s and a also satisfies (3.9), the application of the Leibniz formula to (3.10) yiels

$$\partial^\alpha b(0) = \delta_{0,\alpha}, \quad \text{for all } |\alpha| < s. \tag{3.11}$$

Thus, the sum rules for a and b together with (3.11) imply the sum rules for c.

Let us verify the representation (3.7): by applying the duality (1.26), the interpolatory property (1.48), and

$$\sum_{\tilde\gamma\in\Gamma_M} b(\xi+\gamma+\tilde\gamma) = \sum_{\tilde\gamma\in\Gamma_M} b(\xi+\tilde\gamma), \quad \text{for } \gamma\in\Gamma_M,$$

we obtain

$$\begin{aligned}\sum_{\gamma\in\Gamma_M} (c\overline{a})(\xi+\gamma) &= \sum_{\gamma\in\Gamma_M} \left(b(\xi+\gamma) + \frac{1}{m} - \frac{1}{m}\sum_{\tilde\gamma\in\Gamma_M} b(\xi+\gamma+\tilde\gamma)\right)\overline{a(\xi+\gamma)} \\ &= 1 + \frac{1}{m} - \frac{1}{m}\sum_{\gamma\in\Gamma_M} b(\xi+\gamma).\end{aligned}$$

With (3.8), this yields

$$c(\xi) + 1 - \sum_{\gamma\in\Gamma_M} (c\overline{a})(\xi+\gamma) = b(\xi). \qquad \square$$

So far, we derived $b^{(0)}$ if $a^{(0)}$ is interpolatory. Next, we extend the choice of $b^{(0)}$ to noninterpolatory $a^{(0)}$, i.e., we allow for $a^{(0)} = \widetilde{a}^N$, $N \in \mathbb{N}$ and $\widetilde{a}$ is interpolatory. Then a possible choice of $b^{(0)}$ is addressed in [Der99, JRS99]. In order to avoid complex notation, we only present their results for $m = 2$. Since we are mainly interested in such dilation matrices, this does not mean any far-reaching restriction. For the following lemma, see [Dau92] as well as [Der99, JRS99], and let

$$P_N(x) := \sum_{j=0}^{N-1} \binom{N-1+j}{j} x^j \tag{3.12}$$

denote the *Bézout polynomial* of order N.

Lemma 3.2.3. *Given a dilation matrix M with $m = 2$ and an interpolatory symbol $\widetilde{a}$, then, for $N \in \mathbb{N}$,*

$$b = \overline{\widetilde{a}^N P_{2N}(1 - \widetilde{a})}$$

is a dual symbol of $a = \widetilde{a}^N$.

The application of Lemma 3.2.3 can provide fine wavelet bi-frame constructions. First, we choose a symmetric and interpolatory symbol $\widetilde{a}$ satisfying the sum rules of order s. Fortunately, there is a large variety of candidates in literature such that $a^{(0)} = c^N$ generates a smooth refinable functions φ with small support and much symmetry, cf. [Der99, JRS99] for instance. Note that $a^{(0)}$ satisfies the sum rules of order Ns. Then $b^{(0)}$ can be obtained by Lemma 3.2.3. It also satisfies the sum rules of order Ns, but it usually does not generate a refinable function in $L_2(\mathbb{R}^d)$. Nevertheless, for increasing N, the product $a^{(0)}b^{(0)}$ often generates very smooth refinable functions, see [Der99, JRS99].

Remark 3.2.4. For $m = 2$ and $c = \overline{a}$ in Theorem 3.2.1, the symbol b in (3.7) coincides with b in Lemma 3.2.3 for $N = 1$.

Finally, it should be mentioned that, given some symbol $a^{(0)}$, the choice of a dual symbol $b^{(0)}$ is very complicated if we do not make any assumptions on $a^{(0)}$. It may even happen, that there does not exist any dual symbol. By means of subsymbols A_{γ^*}, $\gamma^* \in \Gamma_M^*$ of $a^{(0)}$, cf. Subsection 1.1.4, one can at least obtain a general condition, which ensures the existence. Introducing z-notation, i.e.,

$$z_i := e^{-2\pi i \xi_i}, \quad \xi_i \in \mathbb{R},\ i = 1, \ldots, d,$$

provides that the subsymbols A_{γ^*} are given by

$$\sum_{k \in \mathbb{Z}^d} a_{Mk+\gamma^*} z^k.$$

Hence, they may also be considered as Laurent polynomials. Then an application of Hilbert's Nullstellensatz to (1.47) provides that $a^{(0)}$ has a dual symbol iff the subsymbols (as Laurent polynomials) do not have any common zeros on $(\mathbb{C} \setminus \{0\})^d$, see [DM97, JRS99] for details.

3.3 Examples

In this section, we present several examples of wavelet bi-frames. Throughout, we apply the following short-hand notation:

$$\begin{aligned}
\Diamond_t &:= \left\{(k_1,\dots,k_d)^\top \in \mathbb{Z}^d : |k_1| + \ldots + |k_d| \leq t\right\},\\
\Box_t &:= \left\{(k_1,\dots,k_d)^\top \in \mathbb{Z}^d : |k_1|,\dots,|k_d| \leq t\right\},\\
z_i &:= e^{-2\pi i \xi_i}, \quad \xi_i \in \mathbb{R},\ i = 1,\dots,d.
\end{aligned}$$

The following smoothness estimates for the $L_\infty(\mathbb{R}^d)$-critical exponent $s_\infty(f)$, see (2.23) as well as Appendix A.1, are accomplished with an implementation of the so-called Villemoes algorithm, see [Vil94] for the theoretical background and [VSo] for the implementation. Hence, for $f \in L_\infty(\mathbb{R}^d)$ with compact support, we compute

$$\sup\left\{s \geq 0 : \int_{\mathbb{R}^d} (1 + \|\xi\|)^s \, |\widehat{f}(\xi)| d\xi < \infty\right\}. \tag{3.13}$$

Then according to (2.25) and some standard textbook on Fourier analyis, (3.13) provides an estimate from below of $s_\infty(f)$.

3.3.1 Wavelet Bi-Frames in Arbitrary Dimensions

First, we choose univariate symbols

$$\check{a}^{(0)}(z) := \left(\frac{1+z}{2}\right)^N \cdot \left(\frac{1+1/z}{2}\right)^N, \tag{3.14}$$

$$\check{b}^{(0)}(z) := \left(\frac{1+z}{2}\right)^N \cdot \left(\frac{1+1/z}{2}\right)^N \cdot P_{2N}\left(1 - \left(\frac{1+z}{2}\right) \cdot \left(\frac{1+1/z}{2}\right)\right), \tag{3.15}$$

where P_{2N} denotes the Bézout polynomial of order $2N$, see (3.12). Since the dyadic univariate symbol

$$\left(\frac{1+z}{2}\right) \cdot \left(\frac{1+1/z}{2}\right) \tag{3.16}$$

is interpolatory, Lemma 3.2.3 yields that $\check{b}^{(0)}$ is dual to $\check{a}^{(0)}$. Moreover, the symbol in (3.16) satisfies the sum rules of order 2. Hence, both $\check{a}^{(0)}$ and $\check{b}^{(0)}$ satisfy the sum rules of order $2N$.

We obtain multivariate symbols by the simple choice

$$a^{(0)}(\xi_1,\dots,\xi_d) := \check{a}^{(0)}(\xi_1), \qquad b^{(0)}(\xi_1,\dots,\xi_d) := \check{b}^{(0)}(\xi_1).$$

We consider these symbols with respect to the dilation matrix M given by (2.27), (2.28). Since

$$\Gamma_M = \left\{0, (\tfrac{1}{2},\dots,\tfrac{1}{2})^\top\right\},$$

the multivariate symbols $a^{(0)}$ and $b^{(0)}$ are also dual to each other. Then Example 1.1.19 provides a symbol family

$$\left\{\left(a^{(0)}, b^{(0)}\right), \left(a^{(1)}, b^{(1)}\right)\right\}$$

satisfying condition (I) on page 34. The wavelet symbols $a^{(1)}$ and $b^{(1)}$ have $2N$ vanishing moments, and according to Corollary 3.1.2, the family

$$\left\{\left(a^{(0)}b^{(0)},a^{(0)}b^{(0)}\right),\left(a^{(1)}b^{(0)},a^{(0)}b^{(1)}\right),\left(a^{(0)}b^{(1)},a^{(1)}b^{(0)}\right),\left(a^{(1)}b^{(1)},a^{(1)}b^{(1)}\right)\right\}$$

also satisfies condition (I). Note that we subsequently verify in Lemma 3.3.1 that the underlying refinable function is contained in $L_2(\mathbb{R}^d)$. Hence, we obtain a wavelet bi-frame. Since the dilation matrix generates the Checkerboard lattice (2.29), we call it *Checkerboard (N)*.

Then $a^{(0)}b^{(0)}$ is interpolatory, and it satisfies the sum rules of order $4N$. Its mask is supported on a straight line in $[-4N+1,4N-1]_\Sigma$. Hence, due to Corollary 2.3.3, $a^{(0)}b^{(0)}$ provides optimal sum rule order. Since the wavelet symbols have at least $2N$ vanishing moments, the wavelet bi-frame provides an optimal approximation order, see Theorem 2.3.1.

The univariate masks of $\check{a}^{(0)}$ and $\check{b}^{(0)}$ are symmetric about a point. Hence, the multivariate symbols are $\mathcal{G}$-symmetric with

$$\mathcal{G} := \{\pm \mathcal{I}_d\}. \tag{3.17}$$

Then, due to Theorem 2.3.5, the three associated wavelets are $\mathcal{G}$-symmetric.

In the following, we verify that $a^{(0)}b^{(0)}$ generates a refinable function with optimal smoothness in the sense of Proposition 2.3.8:

Lemma 3.3.1. *With the choices above, let $\varphi, \widetilde{\varphi}$ be generated by $a^{(0)}$, $b^{(0)}$, respectively. Then the refinable function $\varphi * \widetilde{\varphi}$ of the product $a^{(0)}b^{(0)}$ is given by*

$$\varphi * \widetilde{\varphi}(x) = \prod_{j=1}^{d} \varphi_0 * \widetilde{\varphi}_0\Big(\sum_{i=j}^{d} x_i\Big), \tag{3.18}$$

where $\varphi_0, \widetilde{\varphi}_0$ are generated by $\check{a}^{(0)}$, $\check{b}^{(0)}$, respectively.

Proof. Since $M^d = 2\mathcal{I}_d$, we have

$$M^{\top^{-jd+l}} = 2^{-j} M^{\top^l}, \quad \text{for all } j,l \in \mathbb{N}. \tag{3.19}$$

The infinite product (1.24) and the relation (3.19) provide

$$\begin{aligned}\prod_{j\geq 1} a^{(0)}(M^{\top^{-j}}\xi) &= \prod_{j\geq 1}\prod_{l=0}^{d-1} a^{(0)}(M^{\top^{-jd}}M^{\top^l}\xi)\\ &= \prod_{l=0}^{d-1}\prod_{j\geq 1} a^{(0)}(2^{-j}M^{\top^l}\xi).\end{aligned}$$

By applying the special choices of M and $a^{(0)}$, we obtain

$$\widehat{\varphi}(\xi) = \widehat{\varphi}_0(\xi_d - \xi_{d-1})\cdots\widehat{\varphi}_0(\xi_2-\xi_1)\cdot\widehat{\varphi}_0(\xi_1). \tag{3.20}$$

Following the above lines, we also derive

$$\widehat{\widetilde{\varphi}}(\xi) = \widehat{\widetilde{\varphi}}_0(\xi_d - \xi_{d-1})\cdots\widehat{\widetilde{\varphi}}_0(\xi_2-\xi_1)\cdot\widehat{\widetilde{\varphi}}_0(\xi_1). \tag{3.21}$$

Then a direct calculation of the inverse Fourier transform yields (3.18). □

N	1	2	3	4
mask[1]	$\Diamond_3$	$\Diamond_7$	$\Diamond_{11}$	$\Diamond_{15}$
s_∞	2	3.5	4.7	≥ 5.8
sum rules	4	8	12	16
vm[2]	2	4	6	8
approx.[3]	4	8	12	16

optimality	
sum rules	yes
smoothness	yes
card(wavelets)	–
approx. order	yes
symmetry	yes

Table 3.1: Checkerboard (N)

[1]convex support of the refinable function's mask.
[2]number of vanishing moments which all wavelets inherit.
[3]approximation order of the wavelet bi-frame.

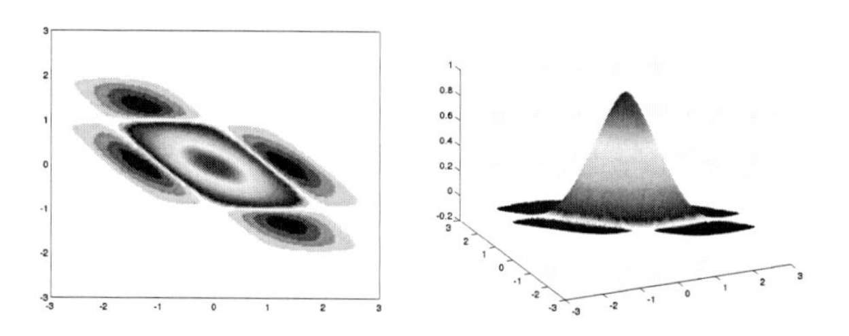

Figure 3.1: The underlying refinable function of Checkerboard (1) for $d = 2$

According to Lemma 3.3.1, the underlying refinable function $\varphi * \widetilde{\varphi}$ inherits the full regularity of the univariate function $\varphi_0 * \widetilde{\varphi}_0$, whose symbol $\check{a}^{(0)}\check{b}^{(0)}$ is interpolatory and satisfies the sum rules of order $4N$. Since the mask of $\check{a}^{(0)}\check{b}^{(0)}$ is supported on $[-4N+1, 4N-1]$, it is the unique interpolatory symbol of Theorem 2.3.2. Thus, due to Proposition 2.3.8, $\varphi * \widetilde{\varphi}$ provides optimal smoothness.

The bi-frame Checkerboard (N) is optimal with respect to almost any optimality criteria in Subsection 2.3.5. It inherits optimal sum rule order, smoothness, approximation order, and symmetry. The only missing condition is the cardinality of the wavelets. We have 3 wavelets, but, according to Lemma 2.3.4, the optimal number is 2, see Table 3.1 for properties and optimality.

The univariate function $\varphi_0 * \widetilde{\varphi}_0$ is essentially the autocorrelation of the underlying refinable function of the N-th orthogonal Daubechies wavelets. It is known that they become arbitrarily smooth by increasing N, cf. [Dau92]. According to (3.18), the smoothness carries over to the multivariate refinable function $\varphi * \widetilde{\varphi}$. Thus, we obtain arbitrarily smooth wavelet bi-frames by increasing N. See Table 3.1 for their properties and Figure 3.1 for $\varphi * \widetilde{\varphi}$ with $d = 2$ and $N = 1$.

Remark 3.3.2. Another bivariate dilation matrix M satisfying $M^2 = 2I_2$ is

$$M = \begin{pmatrix} 0 & 2 \\ 1 & 0 \end{pmatrix}.$$

By applying $\Gamma_M = \{0, (\frac{1}{2}, 0)^\top\}$, we can follow the lines at the beginning of this section. This yields a wavelet bi-frame, where the refinable function $\varphi * \widetilde{\varphi}$ in (3.18) has to be substituted by

$$\varphi * \widetilde{\varphi}(x_1, x_2) = \varphi_0 * \widetilde{\varphi}_0(x_1) \cdot \varphi_0 * \widetilde{\varphi}_0(x_2).$$

Hence, it is separable.

3.3.2 The Quincunx Dilation Matrix

In this section, we address the quincunx dilation matrix

$$M = M_q = \begin{pmatrix} 1 & -1 \\ 1 & 1 \end{pmatrix}. \tag{3.22}$$

Since $m = 2$, our convolution method of Corollary 3.1.2 with $n = 1$ provides a wavelet bi-frame of only 3 wavelets.

Example 3.3.3 (Laplace)**.** The *Laplace symbol*

$$\widetilde{a}(z_1, z_2) := \frac{1}{2}\left(1 + \frac{1}{4}z_1 + \frac{1}{4}\frac{1}{z_1} + \frac{1}{4}z_2 + \frac{1}{4}\frac{1}{z_2}\right) \tag{3.23}$$

is real-valued, interpolatory, satisfies the sum rules of order 2, and it generates a refinable function in $\mathcal{C}^\alpha(\mathbb{R}^2)$, for $\alpha = 0.6$, see [CD93]. For $N_1, N_2 \in \mathbb{N}$, let

$$a := \widetilde{a}^{N_1}, \qquad b := \widetilde{a}^{N_2} P_{N_1+N_2}(1 - \widetilde{a}), \tag{3.24}$$

where P_N is again the Bézout polynomial in (3.12). Then a and b are real-valued, and b is dual to a, see [Dau92, Der99] as well as Lemma 3.2.3 in case of $N_1 = N_2$. Example 1.1.19 provides the family $\{(a, b), (a^{(1)}, b^{(1)})\}$ satisfying condition (I) on page 34. By applying Corollary 3.1.2, the family

$$\left\{(ab, ab), \left(ab^{(1)}, a^{(1)}b\right), \left(a^{(1)}b, ab^{(1)}\right), \left(a^{(1)}b^{(1)}, a^{(1)}b^{(1)}\right)\right\}$$

also satisfies condition (I). The mixed extension principle yields an associated wavelet bi-frame, which we denote by *Laplace (N_1-N_2)*. Observe that a, b, and ab are $\mathcal{G}$-symmetric with

$$\mathcal{G} := \left\{\pm \mathcal{I}_2, \pm \begin{pmatrix} 1 & 0 \\ 0 & -1 \end{pmatrix}, \pm \begin{pmatrix} 0 & -1 \\ 1 & 0 \end{pmatrix}, \pm \begin{pmatrix} 0 & 1 \\ 1 & 0 \end{pmatrix}\right\}. \tag{3.25}$$

The underlying refinable function $\varphi * \widetilde{\varphi}$ is fundamental, and according to Theorem 2.3.5, all wavelets are $\mathcal{G}$-symmetric. See Figure 3.2 and Figure 3.3 for Laplace (1-1).

The optimality results of Laplace (N-N) are summarized in Table 3.2: we have optimality with respect to sum rule order, approximation order, and symmetry. For $N = 1, 2$, properties and a comparison to biorthogonal wavelet bases from [HJ02] are presented in Tables 3.3 and 3.4. The chosen biorthogonal bases are the best known (at least to us), which are comparitive to Laplace (1-1) and (2-2) with respect to smoothness, symmetry, approximation order, and a fundamental primal refinable function. Contrary to the diamond shape of the bi-frame masks, the biorthogonal masks have some square shape $\square_t$. If one does not care about the different shapes, then the biorthogonal mask supports $\square_t$ have to be replaced by $\diamondsuit_{2t}$. Then the superior performance of Laplace (N-N) would even be more significant. In order to guarantee a fair comparison, we respect the different shapes.

$$\left[\begin{array}{ccccccc}
 & & & \frac{-1}{128} & & & \\
 & & \frac{-3}{128} & 0 & \frac{-3}{128} & & \\
 & \frac{-3}{128} & 0 & \frac{39}{128} & 0 & \frac{-3}{128} & \\
\frac{-1}{128} & 0 & \frac{39}{128} & \mathbf{1} & \frac{39}{128} & 0 & \frac{-1}{128} \\
 & \frac{-3}{128} & 0 & \frac{39}{128} & 0 & \frac{-3}{128} & \\
 & & \frac{-3}{128} & 0 & \frac{-3}{128} & & \\
 & & & \frac{-1}{128} & & &
\end{array}\right]$$

(a) ab

$$\left[\begin{array}{ccccc}
 & & \frac{-1}{16} & & \\
 & \frac{-1}{8} & 0 & \frac{-1}{8} & \\
\frac{-1}{16} & \mathbf{0} & \frac{3}{4} & 0 & \frac{-1}{16} \\
 & \frac{-1}{8} & 0 & \frac{-1}{8} & \\
 & & \frac{-1}{16} & &
\end{array}\right]$$

(b) $ab^{(1)}$

$$\left[\begin{array}{ccccccccc}
 & & & & \frac{1}{1024} & & & & \\
 & & & \frac{1}{256} & 0 & \frac{1}{256} & & & \\
 & & \frac{3}{512} & 0 & \frac{-1}{16} & 0 & \frac{3}{512} & & \\
 & \frac{1}{256} & 0 & \frac{-17}{128} & 0 & \frac{-17}{128} & 0 & \frac{1}{256} & \\
\frac{1}{1024} & 0 & \frac{-1}{16} & \mathbf{0} & \frac{185}{256} & 0 & \frac{-1}{16} & 0 & \frac{1}{1024} \\
 & \frac{1}{256} & 0 & \frac{-17}{128} & 0 & \frac{-17}{128} & 0 & \frac{1}{256} & \\
 & & \frac{3}{512} & 0 & \frac{-1}{16} & 0 & \frac{3}{512} & & \\
 & & & \frac{1}{256} & 0 & \frac{1}{256} & & & \\
 & & & & \frac{1}{1024} & & & &
\end{array}\right]$$

(c) $a^{(1)}b$

$$\left[\begin{array}{ccccccc}
 & & & \frac{1}{128} & & & \\
 & & \frac{3}{128} & 0 & \frac{3}{128} & & \\
 & \frac{3}{128} & 0 & \frac{-39}{128} & 0 & \frac{3}{128} & \\
\frac{1}{128} & \mathbf{0} & \frac{-39}{128} & 1 & \frac{-39}{128} & 0 & \frac{1}{128} \\
 & \frac{3}{128} & 0 & \frac{-39}{128} & 0 & \frac{3}{128} & \\
 & & \frac{3}{128} & 0 & \frac{3}{128} & & \\
 & & & \frac{1}{128} & & &
\end{array}\right]$$

(d) $a^{(1)}b^{(1)}$

Figure 3.2: Masks of Laplace (1-1)

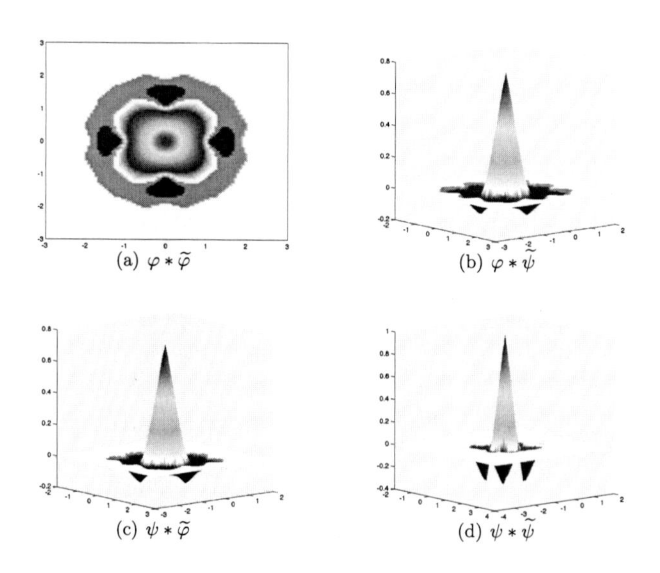

(a) $\varphi * \widetilde{\varphi}$ (b) $\varphi * \widetilde{\psi}$

(c) $\psi * \widetilde{\varphi}$ (d) $\psi * \widetilde{\psi}$

Figure 3.3: Laplace (1-1)

optimality	
sum rules	yes
smoothness	–
card(wavelets)	–
approx. order	yes
symmetry	yes

Table 3.2: Optimality of Laplace (N-N)

First, we address $N = 1$. Laplace (1-1) is smoother than Biorth-A and it provides a higher approximation order. The mask of the underlying refinable function of Laplace (1-1) is supported on the diamond $\Diamond_3$. The dual mask of Biorth-A is supported on the square $\Box_4$. Since

$$\mathrm{card}(\Diamond_3) = 25 < 81 = \mathrm{card}(\Box_4),$$

the bi-frame possesses better properties with smaller mask sizes.

The basis Biorth-B and the bi-frame Laplace (1-1) have similar smoothness. The primal wavelets of Biorth-B have 10 vanishing moments, and in some applications, this may provide a better performance than only 4 vanishing moments of the bi-frame. Nevertheless, both wavelet systems provide the same approximation order 4. Since the primal mask of Biorth-B is supported on $\Box_2$ and $\mathrm{card}(\Diamond_3) = \mathrm{card}(\Box_2)$, it has the same support size as the bi-frame. However, the dual mask of Biorth-B requires a square of size 7. Since $\mathrm{card}(\Box_7)$ equals 225, it is significantly larger than the bi-frame's diamond of size 3. Thus, Laplace (1-1) has similar smoothness and provides the same approximation order with significantly smaller support.

Next, we adress $N = 2$. Laplace (2-2) is two times differentiable, and it provides approximation order 8 with a mask of size $\Diamond_7$. Biorth-C and Biorth-D are less smooth, and they provide less approximation order. Their dual masks are supported on a square of size 5 and 10, respectively. Since $\mathrm{card}(\Diamond_7)$ equals 113, and

$$\mathrm{card}(\Box_5) = 121, \qquad \mathrm{card}(\Box_{10}) = 441,$$

Laplace (2-2) possesses better properties than Biorth-C and Biorth-D with smaller supports.

Smoothness and approximation order of Laplace (2-2) and Biorth-E are comparable. Although the primal wavelets of Biorth-E have 14 vanishing moments, it only provides approximation order 8. The primal mask of Biorth-E has a smaller support size than the bi-frame, i.e., it is supported on a square of size 4, which is smaller than the diamond of size 7, since $\mathrm{card}(\Box_4) = 81 < 113 = \mathrm{card}(\Diamond_7)$. However, the dual mask of Biorth-E is supported on a square of size 11. Since $\mathrm{card}(\Box_{11})$ equals 506, it is much larger. Finally, Laplace (2-2) has similar smoothness and provides the same approximation order with significantly smaller supports.

Finally, let us mention that the results in [JRS99] verify that increasing N_1, N_2 in Laplace (N_1-N_2) yields a family of arbitrarily smooth wavelet bi-frames.

The following example is dedicated to obtaining a three times differentiable wavelet bi-frame for the quincunx dilation matrix. In view of Example 3.3.3, we can procced in two different directions. On the one hand, we can consider Laplace (N_1-N_2) for N_1, N_2

	Laplace (1-1)	Biorth-A[1]		Biorth-B[2]	
	$\varphi * \widetilde{\varphi}$	φ	$\widetilde{\varphi}$	φ	$\widetilde{\varphi}$
mask	$\Diamond_3$	$\Diamond_1$	$\Box_4$	$\Box_2$	$\Box_7$
s_∞	1.3	0.6	0.04	1.4	1.2
sum rules	4	2	6	4	10
vm	2	6	2	10	4
approx.	4	2		4	

Table 3.3: Laplace (1-1) versus biorthogonal bases

[1]best choice in [HJ02] with φ Laplace refinable function with mask (3.23), $\widetilde{\varphi}$ in $L_2(\mathbb{R}^2)$.
[2]best choice in [HJ02] with $\varphi, \widetilde{\varphi} \in \mathcal{C}^1(\mathbb{R}^2)$.

	Laplace (2-2)	Biorth-C[1]		Biorth-D[2]		Biorth-E[3]	
	$\varphi * \widetilde{\varphi}$	φ	$\widetilde{\varphi}$	φ	$\widetilde{\varphi}$	φ	$\widetilde{\varphi}$
mask	$\Diamond_7$	$\Box_2$	$\Box_5$	$\Box_3$	$\Box_{10}$	$\Box_4$	$\Box_{11}$
s_∞	2.5	1.4	0.09	2.2	2.2	2.9	2.2
sum rules	8	4	6	6	14	8	14
vm	4	6	4	14	6	14	8
approx.	8	4		6		8	

Table 3.4: Laplace (2-2) versus biorthogonal bases

[1]best choice in [HJ02] with $\varphi \in \mathcal{C}^1(\mathbb{R}^2)$, $\widetilde{\varphi}$ in $L_2(\mathbb{R}^2)$ (φ coincides with the primal refinable function of Biorth-B).
[2]best choice in [HJ02] with $\varphi, \widetilde{\varphi} \in \mathcal{C}^2(\mathbb{R}^2)$.
[3]best choice in [HJ02] with $\varphi, \widetilde{\varphi} \in \mathcal{C}^2(\mathbb{R}^2)$ and approximation order 8.

$$\begin{bmatrix} & & & \frac{1}{256} & & & \\ & & \frac{-9}{256} & 0 & \frac{-9}{256} & & \\ & \frac{-9}{256} & 0 & \frac{81}{256} & 0 & \frac{-9}{256} & \\ \frac{1}{256} & 0 & \frac{81}{256} & 1 & \frac{81}{256} & 0 & \frac{1}{256} \\ & \frac{-9}{256} & 0 & \frac{81}{256} & 0 & \frac{-9}{256} & \\ & & \frac{-9}{256} & 0 & \frac{-9}{256} & & \\ & & & \frac{1}{256} & & & \end{bmatrix}$$

Figure 3.4: The mask of a in Example 3.3.4

sufficiently large. On the other hand, we can apply the iteration in (3.24) for $N_1 = N_2 = 1$, but we replace the Laplace symbol $\widetilde{a}$ by another symbol with better initial properties.

Example 3.3.4 (Three times differentiable)**.** The symbol a given by the mask in Figure 3.4 is interpolatory and satisfies the sum rules of order 4. It generates a fundamental refinable function φ contained in the Hölder class $\mathcal{C}^{\alpha}(\mathbb{R}^2)$, for $\alpha = 1.5$, see [DGM99]. By Theorem 3.2.1,

$$b(\xi) := a(\xi) + 1 - |a(\xi)|^2 - |a(\xi+\gamma)|^2 \tag{3.26}$$

is dual to a. The symbols a, b, and ab are $\mathcal{G}$-symmetric with $\mathcal{G}$ as in (3.25). Let φ and $\widetilde{\varphi}$ denote the refinable functions of a and b, respectively. Then Example 1.1.19 and Corollary 3.1.2 provide a symbol family

$$\left\{(ab, ab), \left(ab^{(1)}, a^{(1)}b\right), \left(a^{(1)}b, ab^{(1)}\right), \left(a^{(1)}b^{(1)}, a^{(1)}b^{(1)}\right)\right\}$$

inducing a wavelet bi-frame

$$X(\{\varphi * \widetilde{\psi}, \psi * \widetilde{\varphi}, \psi * \widetilde{\psi}\}), \quad X(\{\psi * \widetilde{\varphi}, \varphi * \widetilde{\psi}, \psi * \widetilde{\psi}\})$$

with fundamental refinable function $\varphi * \widetilde{\varphi}$. We call the bi-frame *DGM*. According to Lemma 2.3.5, all wavelets are $\mathcal{G}$-symmetric, see Figure 3.5. However, we only achieve the estimate $s_\infty(\varphi * \widetilde{\varphi}) \approx 2.9$. Laplace (3-2) and Laplace (2-3) provide three times differentiable wavelets with the same mask support size $\Diamond_9$, see Table 3.5 for the comparison of DGM, Laplace (3-2), and the smoothest comparative biorthogonal basis Biorth-F from [HJ02]. Biorth-F does not reach $s_\infty = 3$, and its dual mask is supported on a square of size 12. Since $\operatorname{card}(\Box_{12}) = 625$ is much larger than the bi-frame's $\operatorname{card}(\Diamond_9) = 181$, we can establish that both bi-frames perform better than the biorthogonal basis.

Finally, optimality results are summarized in Table 3.6. Both bi-frames provide optimal symmetry. DGM also provides optimal approximation order, but it is not optimal with respect to sum rules. Laplace (3-2) has optimal sum rule order, but it does not provide optimal approximation order.

3.3.3 A Bivariate Box Spline Wavelet Bi-Frame

In this section, we deal with a three-direction box spline. We choose $M = 2\mathcal{I}_2$, then let

$$a(z) := \left(\frac{1+z_1}{2}\right) \cdot \left(\frac{1+z_2}{2}\right) \cdot \left(\frac{1+z_1z_2}{2}\right) \cdot \left(\frac{1}{z_1z_2}\right). \tag{3.27}$$

Hence, a is essentially the symbol of a three-direction box spline with direction matrix

$$Y_3 = \begin{pmatrix} 1 & 0 & 1 \\ 0 & 1 & 1 \end{pmatrix},$$

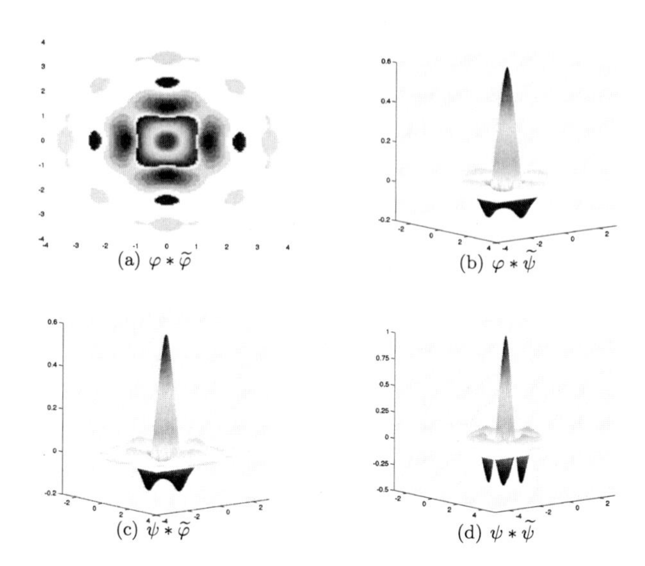

(a) $\varphi * \widetilde{\varphi}$ (b) $\varphi * \widetilde{\psi}$ (c) $\psi * \widetilde{\varphi}$ (d) $\psi * \widetilde{\psi}$

Figure 3.5: DGM bi-frame

	DGM	Laplace (3-2)	Biorth-F[1]	
	$\varphi * \widetilde{\varphi}$	$\varphi * \widetilde{\varphi}$	φ	$\widetilde{\varphi}$
mask	$\Diamond_9$	$\Diamond_9$	$\Box_4$	$\Box_{12}$
s_∞	2.9	3.0	2.9	2.8
sum rules	8	10	8	16
vm	4	4	16	8
approx.	8	8	8	

Table 3.5: DGM versus Laplace (3-2) versus biorthogonal basis

[1] φ coincides with the primal refinable function in Biorth-E.

optimality		
	DGM	Laplace (3-2)
sum rules	–	yes
smoothness	–	–
card(wavelets)	–	–
approx. order	yes	–
symmetry	yes	yes

Table 3.6: Optimality of DGM and Laplace (3-2)

see Lemma 1.1.16. The factor $\frac{1}{z_1 z_2}$ only yields a shift of the box spline, and it provides that a is real-valued. From the results in [RS97a], we obtain a real-valued symbol $\widetilde{b}$ such that $b = a\widetilde{b}$ is dual to a. Since a is interpolatory, Lemma 1.1.20 is applicable and provides a family of start symbols

$$\left\{\left(a^{(\mu)}, b^{(\mu)}\right) : \mu = 0, \ldots, 4\right\},$$

where $a = a^{(0)}$ and $b = b^{(0)}$. Then Corollary 3.1.2 yields a wavelet bi-frame of 15 wavelets, which we call *Box Spline (1)*. The underlying refinable function is fundamental, and its mask is $\mathcal{G}$-symmetric with

$$\mathcal{G} := \left\{\pm \mathcal{I}_2, \pm \begin{pmatrix} 0 & 1 \\ 1 & 0 \end{pmatrix}, \pm \begin{pmatrix} 1 & 0 \\ 1 & -1 \end{pmatrix}, \pm \begin{pmatrix} 0 & 1 \\ -1 & 1 \end{pmatrix}, \pm \begin{pmatrix} 1 & -1 \\ 1 & 0 \end{pmatrix}, \pm \begin{pmatrix} -1 & 1 \\ 0 & 1 \end{pmatrix}\right\}, \quad (3.28)$$

see Figure 3.6. According to the symmetries of the wavelet masks, Theorem 2.3.5 yields symmetry of the wavelets with respect to different subgroups $\mathcal{G}_i \subset \mathcal{G}$, see also Figures 3.7 and 3.8:

$$\mathcal{G}_1 := \left\{\pm \mathcal{I}_2, \pm \begin{pmatrix} 0 & 1 \\ 1 & 0 \end{pmatrix}\right\}, \qquad \varphi * \widetilde{\psi}^{(3)}, \psi^{(3)} * \widetilde{\varphi}, \psi^{(3)} * \widetilde{\psi}^{(3)} \text{ are } \mathcal{G}_1\text{-symmetric.}$$

$$\mathcal{G}_2 := \left\{\pm \mathcal{I}_2, \pm \begin{pmatrix} -1 & 1 \\ 0 & 1 \end{pmatrix}\right\}, \qquad \varphi * \widetilde{\psi}^{(1)}, \psi^{(1)} * \widetilde{\varphi}, \psi^{(1)} * \widetilde{\psi}^{(1)} \text{ are } \mathcal{G}_2\text{-symmetric.}$$

$$\mathcal{G}_3 := \left\{\pm \mathcal{I}_2, \pm \begin{pmatrix} 1 & 0 \\ 1 & -1 \end{pmatrix}\right\}, \qquad \varphi * \widetilde{\psi}^{(2)}, \psi^{(2)} * \widetilde{\varphi}, \psi^{(2)} * \widetilde{\psi}^{(2)} \text{ are } \mathcal{G}_3\text{-symmetric.}$$

$$\mathcal{G}_4 := \{\pm \mathcal{I}_2\}, \qquad \text{all other wavelets are } \mathcal{G}_4\text{-symmetric.}$$

Since the wavelets lose symmetry with respect to the underlying refinable function, the bi-frame does not have optimal symmetry. Nevertheless, it provides optimal approximation order, sum rule order, and smoothness, see Table 3.8.

A comparison to biorthogonal bases from [CHR00, HR02] is presented in Table 3.7. The mask of our bi-frame's refinable function is supported on a square of size 3. The $L_\infty(\mathbb{R}^2)$-critical exponent equals 2, and it provides approximation order 4. The dual masks of the biorthogonal bases Biorth-G and Biorth-H have a much larger support since they require a square of size 6. Moreover, they are less smooth and they do not provide a higher approximation order than the bi-frame. Thus, Box Spline (1) performs better than comparable biorthogonal bases.

It should be mentioned that interchanging the application of Corollary 3.1.2 by part (b) of Theorem 3.1.1 yields a wavelet bi-frame of 6 wavelets, but the refinable function is no longer fundamental.

$$\begin{bmatrix} & & & \frac{-1}{64} & \frac{-3}{64} & \frac{-3}{64} & \frac{-1}{64} \\ & & \frac{-3}{64} & 0 & \frac{3}{32} & 0 & \frac{-3}{64} \\ & \frac{-3}{64} & \frac{3}{32} & \frac{33}{64} & \frac{33}{64} & \frac{3}{32} & \frac{-3}{64} \\ \frac{-1}{64} & 0 & \frac{33}{64} & 1 & \frac{33}{64} & 0 & \frac{-1}{64} \\ \frac{-3}{64} & \frac{3}{32} & \frac{33}{64} & \frac{33}{64} & \frac{3}{32} & \frac{-3}{64} & \\ \frac{-3}{64} & 0 & \frac{3}{32} & 0 & \frac{-3}{64} & & \\ \frac{-1}{64} & \frac{-3}{64} & \frac{-3}{64} & \frac{-1}{64} & & & \end{bmatrix}$$

(a) $a^{(0)}b^{(0)}$

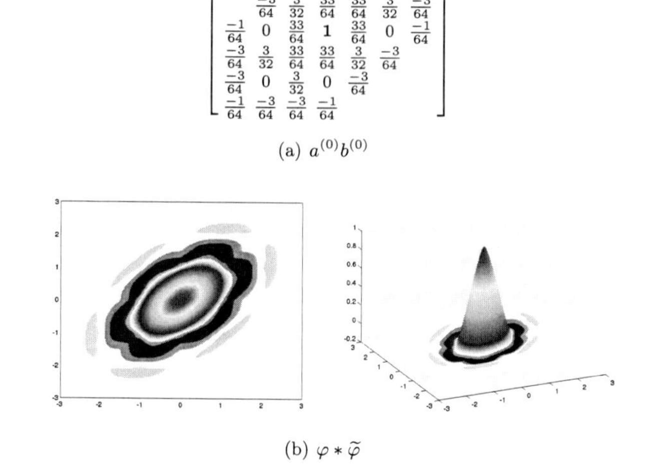

(b) $\varphi * \widetilde{\varphi}$

Figure 3.6: Underlying refinable function with its mask of Box Spline (1)

	Box Spline (1)	Biorth-G[1]		Biorth-H[2]	
	$\varphi * \widetilde{\varphi}$	φ	$\widetilde{\varphi}$	φ	$\widetilde{\varphi}$
mask	$\square_3$	$\square_1$	$\square_6$	$\square_3$	$\square_6$
s_∞	2	1	0.6[3]	1.4[3]	0.07[3]
sum rules	4	2	6	4	4
vm	2	6	2	4	4
approx.	4	2		4	

Table 3.7: Box Spline (1) versus biorthogonal bases

[1] φ is the three-direction box spline with multiplicity 1. $\widetilde{\varphi}$ is constructed in [HR02].
[2] φ (butterfly mask) and $\widetilde{\varphi}$ are given in [CHR00].
[3] obtained by Sobolev embedding.

optimality	
sum rules	yes
smoothness	yes
card(wavelets)	–
approx. order	yes
symmetry	–

Table 3.8: Optimality of the Box Spline (1)

$$\begin{bmatrix} & \frac{-1}{16} & \frac{1}{16} & \frac{1}{16} & \frac{-1}{16} \\ \frac{-1}{16} & \mathbf{0} & \frac{1}{8} & 0 & \frac{-1}{16} \\ \frac{-1}{16} & \frac{1}{16} & \frac{1}{16} & \frac{-1}{16} & \end{bmatrix}$$

(a) $a^{(0)}b^{(1)}$

$$\begin{bmatrix} & \frac{-1}{16} & \frac{-1}{16} \\ \frac{-1}{16} & 0 & \frac{1}{16} \\ \frac{1}{16} & \frac{1}{8} & \frac{1}{16} \\ \frac{1}{16} & \mathbf{0} & \frac{-1}{16} \\ \frac{-1}{16} & \frac{-1}{16} & \end{bmatrix}$$

(b) $a^{(0)}b^{(2)}$

$$\begin{bmatrix} & & & \frac{-1}{16} & \frac{-1}{16} \\ & & \frac{1}{16} & 0 & \frac{-1}{16} \\ & \frac{1}{16} & \frac{1}{8} & \frac{1}{16} & \\ \frac{-1}{16} & \mathbf{0} & \frac{1}{16} & & \\ \frac{-1}{16} & \frac{-1}{16} & & & \end{bmatrix}$$

(c) $a^{(0)}b^{(3)}$

$$\begin{bmatrix} & & & \frac{-1}{64} & \frac{1}{64} & \frac{1}{64} & \frac{-1}{64} \\ & & \frac{-1}{64} & 0 & \frac{1}{32} & 0 & \frac{-1}{64} \\ & \frac{3}{64} & \frac{-1}{16} & \frac{1}{64} & \frac{1}{64} & \frac{-1}{16} & \frac{3}{64} \\ \frac{3}{64} & \mathbf{0} & \frac{-35}{64} & 1 & \frac{-35}{64} & 0 & \frac{3}{64} \\ \frac{3}{64} & \frac{-1}{16} & \frac{1}{64} & \frac{1}{64} & \frac{-1}{16} & \frac{3}{64} & \\ \frac{-1}{64} & 0 & \frac{1}{32} & 0 & \frac{-1}{64} & & \\ \frac{-1}{64} & \frac{1}{64} & \frac{1}{64} & \frac{-1}{64} & & & \end{bmatrix}$$

(d) $a^{(1)}b^{(1)}$

$$\begin{bmatrix} & & & \frac{3}{64} & \frac{3}{64} & \frac{-1}{64} & \frac{-1}{64} \\ & & \frac{3}{64} & 0 & \frac{-1}{16} & 0 & \frac{1}{64} \\ & \frac{-1}{64} & \frac{-1}{16} & \frac{-35}{64} & \frac{1}{64} & \frac{1}{32} & \frac{1}{64} \\ \frac{-1}{64} & 0 & \frac{1}{64} & 1 & \frac{1}{64} & 0 & \frac{-1}{64} \\ \frac{1}{64} & \frac{1}{32} & \frac{1}{64} & \frac{-35}{64} & \frac{-1}{16} & \frac{-1}{64} & \\ \frac{1}{64} & 0 & \frac{-1}{16} & \mathbf{0} & \frac{3}{64} & & \\ \frac{-1}{64} & \frac{-1}{64} & \frac{3}{64} & \frac{3}{64} & & & \end{bmatrix}$$

(e) $a^{(2)}b^{(2)}$

$$\begin{bmatrix} & & & \frac{-1}{64} & \frac{-1}{64} & \frac{3}{64} & \frac{3}{64} \\ & & \frac{1}{64} & 0 & \frac{-1}{16} & 0 & \frac{3}{64} \\ & \frac{1}{64} & \frac{1}{32} & \frac{1}{64} & \frac{-35}{64} & \frac{-1}{16} & \frac{-1}{64} \\ \frac{-1}{64} & 0 & \frac{1}{64} & 1 & \frac{1}{64} & 0 & \frac{-1}{64} \\ \frac{-1}{64} & \frac{-1}{16} & \frac{-35}{64} & \frac{1}{64} & \frac{1}{32} & \frac{1}{64} & \\ \frac{3}{64} & \mathbf{0} & \frac{-1}{16} & 0 & \frac{1}{64} & & \\ \frac{3}{64} & \frac{3}{64} & \frac{-1}{64} & \frac{-1}{64} & & & \end{bmatrix}$$

(f) $a^{(3)}b^{(3)}$

$$\begin{bmatrix} & & & \frac{3}{64} & \frac{-3}{64} & \frac{-1}{16} & \frac{1}{16} & \frac{1}{64} & \frac{-1}{64} \\ & & \frac{3}{64} & 0 & \frac{-7}{64} & 0 & \frac{5}{64} & 0 & \frac{-1}{64} \\ & \frac{-1}{64} & \frac{1}{16} & \frac{-31}{64} & \frac{7}{8} & \frac{-31}{64} & \frac{1}{16} & \frac{-1}{64} & \\ \frac{-1}{64} & 0 & \frac{5}{64} & \mathbf{0} & \frac{-7}{64} & 0 & \frac{3}{64} & & \\ \frac{-1}{64} & \frac{1}{64} & \frac{1}{16} & \frac{-1}{16} & \frac{-3}{64} & \frac{3}{64} & & & \end{bmatrix}$$

(g) $a^{(2)}b^{(1)}$

$$\begin{bmatrix} & \frac{-1}{64} & \frac{1}{64} & \frac{1}{16} & \frac{-1}{16} & \frac{-3}{64} & \frac{3}{64} \\ \frac{-1}{64} & 0 & \frac{5}{64} & 0 & \frac{-7}{64} & 0 & \frac{3}{64} \\ \frac{-1}{64} & \frac{1}{16} & \frac{-31}{64} & \frac{7}{8} & \frac{-31}{64} & \frac{1}{16} & \frac{-1}{64} \\ \frac{3}{64} & \mathbf{0} & \frac{-7}{64} & 0 & \frac{5}{64} & 0 & \frac{-1}{64} \\ \frac{3}{64} & \frac{-3}{64} & \frac{-1}{16} & \frac{1}{16} & \frac{1}{64} & \frac{-1}{64} & \end{bmatrix}$$

(h) $a^{(3)}b^{(1)}$

$$\begin{bmatrix} & \frac{-1}{64} & \frac{-1}{64} & \frac{3}{64} & \frac{3}{64} \\ \frac{-1}{64} & 0 & \frac{1}{16} & 0 & \frac{-3}{64} \\ \frac{1}{64} & \frac{5}{64} & \frac{-31}{64} & \frac{-7}{64} & \frac{-1}{16} \\ \frac{1}{16} & 0 & \frac{7}{8} & 0 & \frac{1}{16} \\ \frac{-1}{16} & \frac{-7}{64} & \frac{-31}{64} & \frac{5}{64} & \frac{1}{64} \\ \frac{-3}{64} & \mathbf{0} & \frac{1}{16} & 0 & \frac{-1}{64} \\ \frac{3}{64} & \frac{3}{64} & \frac{-1}{64} & \frac{-1}{64} & \end{bmatrix}$$

(i) $a^{(3)}b^{(2)}$

$$\begin{bmatrix} & & & \frac{-1}{64} & \frac{-1}{64} \\ & & \frac{-1}{64} & 0 & \frac{1}{64} \\ & \frac{3}{64} & \frac{1}{16} & \frac{5}{64} & \frac{1}{16} \\ \frac{3}{64} & 0 & \frac{-31}{64} & 0 & \frac{-1}{16} \\ \frac{-3}{64} & \frac{-7}{64} & \frac{7}{8} & \frac{-7}{64} & \frac{-3}{64} \\ \frac{-1}{16} & \mathbf{0} & \frac{-31}{64} & 0 & \frac{3}{64} \\ \frac{1}{16} & \frac{5}{64} & \frac{1}{16} & \frac{3}{64} & \\ \frac{1}{64} & 0 & \frac{-1}{64} & & \\ \frac{-1}{64} & \frac{-1}{64} & & & \end{bmatrix}$$

(j) $a^{(1)}b^{(2)}$

$$\begin{bmatrix} & & & & & \frac{-1}{64} & \frac{-1}{64} \\ & & & & \frac{1}{64} & 0 & \frac{-1}{64} \\ & & & \frac{1}{16} & \frac{5}{64} & \frac{1}{16} & \frac{3}{64} \\ & & \frac{-1}{16} & 0 & \frac{-31}{64} & 0 & \frac{3}{64} \\ & \frac{-3}{64} & \frac{-7}{64} & \frac{7}{8} & \frac{-7}{64} & \frac{-3}{64} & \\ \frac{3}{64} & \mathbf{0} & \frac{-31}{64} & 0 & \frac{-1}{16} & & \\ \frac{3}{64} & \frac{1}{16} & \frac{5}{64} & \frac{1}{16} & & & \\ \frac{-1}{64} & 0 & \frac{1}{64} & & & & \\ \frac{-1}{64} & \frac{-1}{64} & & & & & \end{bmatrix}$$

(k) $a^{(1)}b^{(3)}$

$$\begin{bmatrix} & & & & & \frac{3}{64} & \frac{3}{64} & \frac{-1}{64} & \frac{-1}{64} \\ & & & & \frac{-3}{64} & 0 & \frac{1}{16} & 0 & \frac{-1}{64} \\ & & & \frac{-1}{16} & \frac{-7}{64} & \frac{-31}{64} & \frac{5}{64} & \frac{1}{64} & \\ & & \frac{1}{16} & 0 & \frac{7}{8} & 0 & \frac{1}{16} & & \\ & \frac{1}{64} & \frac{5}{64} & \frac{-31}{64} & \frac{-7}{64} & \frac{-1}{16} & & & \\ \frac{-1}{64} & 0 & \frac{1}{16} & \mathbf{0} & \frac{-3}{64} & & & & \\ \frac{-1}{64} & \frac{-1}{64} & \frac{3}{64} & \frac{3}{64} & & & & & \end{bmatrix}$$

(l) $a^{(2)}b^{(3)}$

Figure 3.7: Masks of Box Spline (1), part I

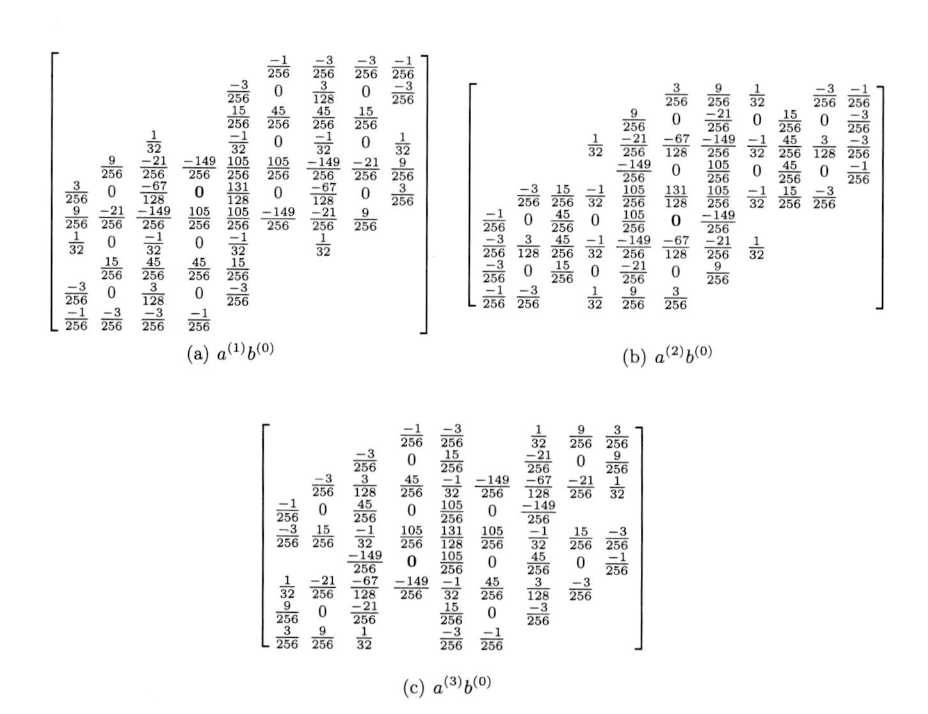

(a) $a^{(1)}b^{(0)}$ (b) $a^{(2)}b^{(0)}$ (c) $a^{(3)}b^{(0)}$

Figure 3.8: Masks of Box Spline (1), part II

Chapter 4

Wavelet Bi-Frames with Few Wavelets

Provided that $m = 2$, the wavelet bi-frames in Chapter 3 are generated by three wavelets. This yields a sufficiently small complexity of the fast wavelet frame transform. Nevertheless, since the refinable functions in the examples are fundamental, Lemma 2.3.4 yields that the optimal number of wavelets is two. The present chapter is dedicated to minimizing the number of wavelets from three to two.

In order to reduce the number of wavelets, we try to adapt the univariate dyadic ideas in [CHS02, DHRS03]. They are based on the so-called mixed oblique extension principle generalizing the mixed extension principle in Theorem 2.2.1. The univariate approach can be splitted into two steps. Step one requires the solution of a generalization of the modulation matrix completion problem as discussed in Subsection 1.1.4. In Step two, the univariate approach requires a factorization of some trigonometric polynomial into two trigonometric polynomials satisfying certain zero conditions.

The present chapter is organized as follows: first, we recall the mixed oblique extension principle. Wavelet bi-frames from the oblique principle provide a fast oblique wavelet frame transform. In comparison to the fast wavelet frame transform of the mixed extension principle, it requires an additional convolution at the beginning of the reconstruction process as well as a deconvolution at its end. Then we adapt the univariate dyadic ideas in [CHS02, DHRS03] to our multivariate setting with general dilation matrices. In order to solve the generalization of the modulation matrix completion problem, we reformulate the problem in terms of polyphases. Then we complete the polyphase matrices, which also provides a completion of the original modulation matrices for Step one. Next, we address Step two. In the multivariate setting, such factorizations of trigonometric polynomials are generally impossible. We circumvent this difficulty by allowing for a sum of products. Then we derive conditions, which ensure that we can find a convenient sum of products for the application of the modified Step two. It should be mentioned that these conditions still provide sufficient flexibility. Then in comparison to the method of the previous chapter, the number of wavelets is reduced from $m^2 - 1$ in Chapter 3 to $3m - 4$ wavelets. Finally, we obtain wavelet bi-frames from the refinable functions of the examples in Chapter 3. Then the bi-frames in the present chapter possess similar properties with significantly fewer wavelets.

4.1 An Oblique Wavelet Bi-Frame Construction

4.1.1 The Mixed Oblique Extension Principle

In this section, we recall a generalization of the mixed extension principle. It is based on a more general version of condition (I) on page 34. Given a symbol θ, we say the symbol family $\{(a^{(\mu)}, b^{(\mu)}) : \mu = 0, \dots, n\}$ satisfies *condition (II)* with respect to θ if the following holds:

(II-a) $a^{(0)}(0) = b^{(0)}(0) = \theta(0) = 1$.

(II-b) $a^{(\mu)}(0) = b^{(\mu)}(0) = 0$, for all $1 \leq \mu \leq n$,

(II-c) for all $\gamma \in \Gamma_M$, $\xi \in \mathbb{R}^d$,

$$\overline{a^{(0)}(\xi + \gamma)} b^{(0)}(\xi) \theta(M^\top \xi) + \sum_{\mu=1}^{n} \overline{a^{(\mu)}(\xi + \gamma)} b^{(\mu)}(\xi) = \delta_{0,\gamma} \theta(\xi). \tag{4.1}$$

The next theorem is called the *mixed oblique extension principle*, see [CHS02, DHRS03] as well as [DH00, Han03b]:

Theorem 4.1.1 (MOEP). *Let the symbol family* $\{(a^{(\mu)}, b^{(\mu)}) : \mu = 0, \dots, n\}$ *satisfy* condition (II) *with respect to* θ. *Moreover, let* $a^{(0)}$, $b^{(0)}$ *generate* $\varphi, \widetilde{\varphi} \in L_2(\mathbb{R}^d)$, *respectively. For* $\mu = 1, \dots, n$, *define*

$$\psi^{(\mu)}(\xi) := \sum_{k \in \mathbb{Z}^d} a_k^{(\mu)} \varphi(M\xi - k) \quad \text{and} \quad \widetilde{\psi}^{(\mu)}(\xi) := \sum_{k \in \mathbb{Z}^d} b_k^{(\mu)} \widetilde{\varphi}(M\xi - k) \,. \tag{4.2}$$

Then $X(\{\psi^{(1)}, \dots, \psi^{(n)}\})$, $X(\{\widetilde{\psi}^{(1)}, \dots, \widetilde{\psi}^{(n)}\})$ *is a wavelet bi-frame.*

Contrary to the mixed extension principle, its oblique counterpart allows for $\theta \not\equiv 1$. Hence, it provides additional flexibility. In the present chapter, we use this flexibility for the construction of wavelet bi-frames with fewer wavelets than those in Chapter 3.

First, we discuss the effects of the more general construction principle to the wavelet transform. According to [DHRS03], the mixed oblique extension principle gives rise to the *fast oblique wavelet frame transform*, see Algorithm 3. The decomposition is identical to the fast wavelet frame transform in Algorithm 2. Again, we expect the input sequence $H_0^{(0)}$ either to be the coefficient sequence in (2.18) or directly the inner product in (2.21). Due to (II-c), the reconstruction process involves the symbol θ. It requires the deconvolution of θ at the end of the reconstruction (it may be accomplished similar to the deconvolution of $[\widehat{\varphi}, \widehat{\widetilde{\varphi}}]$ in the fast wavelet frame transform, see Subsection 2.3.2). The deconvolution of θ has a sharpening effect, see [DHRS03]. If perfect reconstruction is not required, then the transform can be applied without the deconvolution. This still provides a blurry version of the original signal. It can be a feature since measured data is generally noisy and it often requires some pre-smoothening. Then abandoning the deconvolution makes such pre-processing unnecessary.

Finally, we address the question about the minimal number of wavelets. According to Lemma 2.3.4, we have already derived some optimality constraints with respect to the mixed extension principle. In the dyadic univariate setting, this lemma still holds with respect to its oblique counterpart, see [DHRS03]:

Theorem 4.1.2. *In the dyadic univariate setting, let the symbol family*

$$\left\{ \left(a^{(0)}, a^{(0)}\right), \left(a^{(1)}, b^{(1)}\right), \dots, \left(a^{(n)}, b^{(n)}\right) \right\}$$

satisfy condition (II) *on page 68 with respect to* θ. *If* $a^{(0)}$ *generates a fundamental refinable function, then* $n \geq 2$.

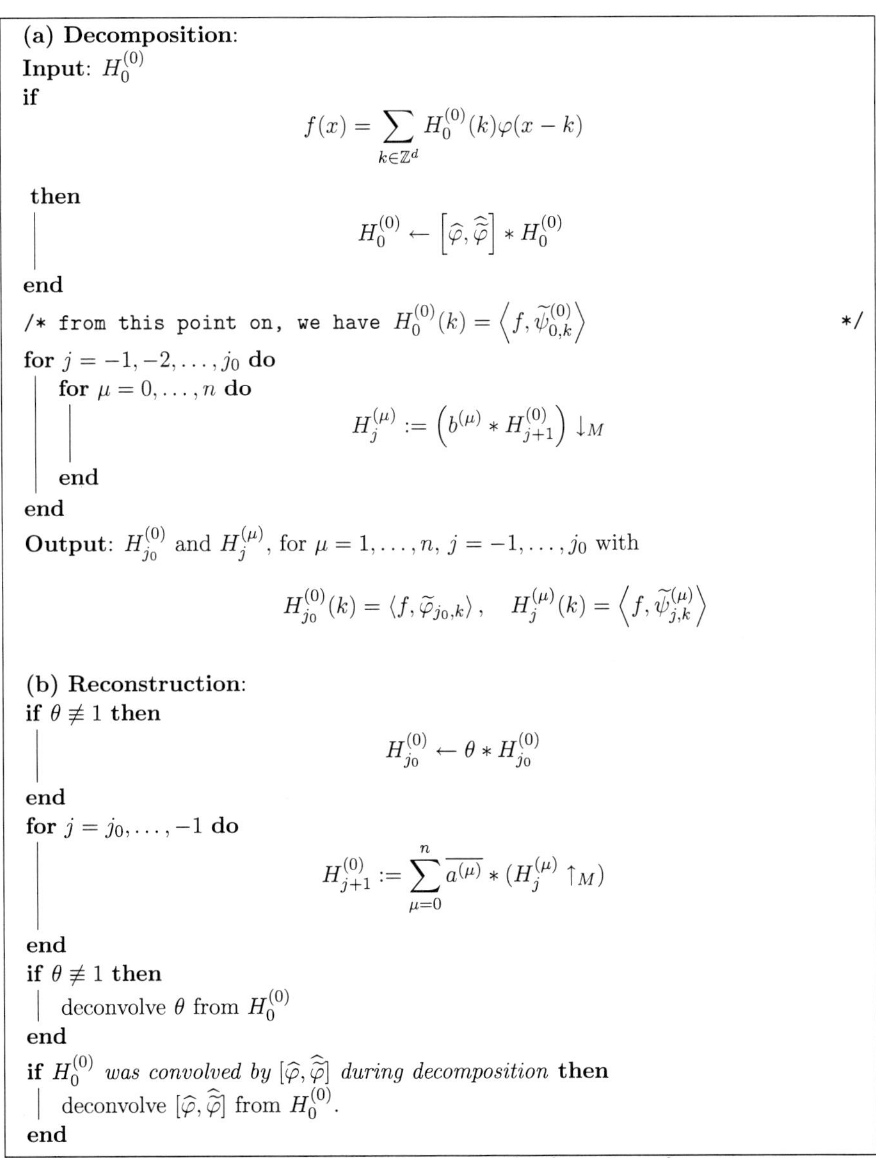

(a) Decomposition:

Input: $H_0^{(0)}$

if

$$f(x) = \sum_{k \in \mathbb{Z}^d} H_0^{(0)}(k) \varphi(x-k)$$

then

$$H_0^{(0)} \leftarrow \left[\widehat{\varphi}, \widehat{\widetilde{\varphi}}\right] * H_0^{(0)}$$

end

`/* from this point on, we have` $H_0^{(0)}(k) = \left\langle f, \widetilde{\psi}_{0,k}^{(0)} \right\rangle$ `*/`

for $j = -1, -2, \ldots, j_0$ **do**

  for $\mu = 0, \ldots, n$ **do**

$$H_j^{(\mu)} := \left(b^{(\mu)} * H_{j+1}^{(0)}\right) \downarrow_M$$

  end

end

Output: $H_{j_0}^{(0)}$ and $H_j^{(\mu)}$, for $\mu = 1, \ldots, n$, $j = -1, \ldots, j_0$ with

$$H_{j_0}^{(0)}(k) = \langle f, \widetilde{\varphi}_{j_0,k} \rangle, \quad H_j^{(\mu)}(k) = \left\langle f, \widetilde{\psi}_{j,k}^{(\mu)} \right\rangle$$

(b) Reconstruction:

if $\theta \not\equiv 1$ **then**

$$H_{j_0}^{(0)} \leftarrow \theta * H_{j_0}^{(0)}$$

end

for $j = j_0, \ldots, -1$ **do**

$$H_{j+1}^{(0)} := \sum_{\mu=0}^{n} \overline{a^{(\mu)}} * (H_j^{(\mu)} \uparrow_M)$$

end

if $\theta \not\equiv 1$ **then**

  deconvolve θ from $H_0^{(0)}$

end

if $H_0^{(0)}$ *was convolved by* $[\widehat{\varphi}, \widehat{\widetilde{\varphi}}]$ *during decomposition* **then**

  deconvolve $[\widehat{\varphi}, \widehat{\widetilde{\varphi}}]$ from $H_0^{(0)}$.

end

Algorithm 3: The fast oblique wavelet frame transform

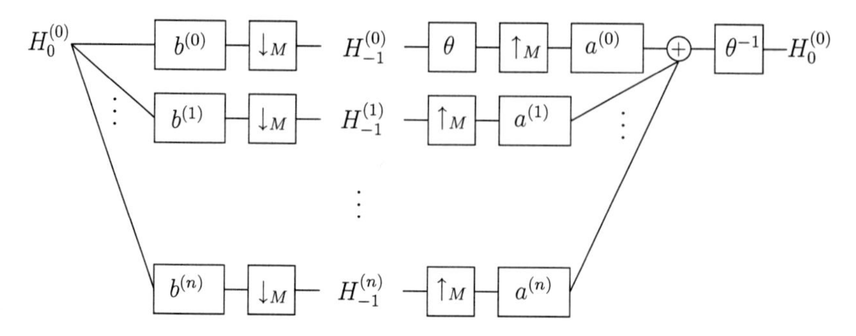

Figure 4.1: The filter bank scheme of the fast oblique wavelet frame transform

In general, the multivariate setting with $m = 2$ does not allow for more flexibility than the dyadic univariate one. Thus, we conjecture that Theorem 4.1.2 still holds in this more general situation. However, so far, we have no proof. The optimality statements with respect to the number of wavelets in the examples of Section 4.3 refer to this conjecture. In other words, the number is optimal if we would have derived the bi-frame from the mixed extension principle.

4.1.2 A General Construction Idea

In the following theorem, we adapt the dyadic univariate construction ideas in [CHS02, DHRS03] to the multivariate setting with general dilation matrices. See also Corollary 14.8.3 in [Chr03] for the univariate ideas.

Theorem 4.1.3. *Let symbols $a^{(0)}$ and $b^{(0)}$ satisfy the sum rules of order* 1. *Moreover, let $a^{(0)}(0) = b^{(0)}(0) = 1$, and denote*

$$\theta(\xi) := \sum_{\gamma\in\Gamma_M} \left(\overline{a^{(0)}}b^{(0)}\right)(\xi+\gamma), \qquad \eta := 1-\theta. \tag{4.3}$$

Assume that

(S1) *there are additional symbols $a^{(\mu)}$, $b^{(\mu)}$, $\mu = 1,\dots,m-1$, such that*

$$\sum_{\mu=0}^{m-1} \overline{a^{(\mu)}(\xi+\gamma)}b^{(\mu)}(\xi) = \delta_{0,\gamma}\theta(\xi),$$

(S1*) *all symbols $a^{(\mu)}$, $b^{(\mu)}$, $\mu = 1,\dots,m-1$, have $s_1 \geq 1$ vanishing moments,*

(S2) *there are additional symbols $\eta^{(\nu)}$, $\widetilde{\eta}^{(\nu)}$, $\nu = 1,\dots,n$, such that*

$$\eta = \sum_{\nu=1}^{n} \overline{\eta^{(\nu)}}\widetilde{\eta}^{(\nu)},$$

(S2*) *all symbols $\eta^{(\nu)}$, $\widetilde{\eta}^{(\nu)}$, $\nu = 1,\dots,n$, have $s_2 \geq 1$ vanishing moments.*

For $\nu = 1, \dots, n$, let

$$a^{(m-1+\nu)}(\xi) := \eta^{(\nu)}(M^\top \xi) a^{(0)}(\xi), \qquad b^{(m-1+\nu)}(\xi) := \widetilde{\eta}^{(\nu)}(M^\top \xi) b^{(0)}(\xi).$$

Then the collection $\{(a^{(\mu)}, b^{(\mu)}) : \mu = 0, \dots, m-1+n\}$ satisfies condition (II) *on page 68 with respect to θ. All wavelet symbols have at least* $\min\{s_1, s_2\}$ *vanishing moments.*

The application of Theorem 4.1.3 in the next sections works as follows: first, we choose $a^{(0)}$ and $b^{(0)}$, then we derive all other symbols by some canonical choice. In the section 4.2, we impose restrictions on the symbol $a^{(0)}$ such that we obtain a simple and concrete construction algorithm.

Remark 4.1.4. If $\theta \equiv 1$, then (4.3) requires that $b^{(0)}$ is dual to $a^{(0)}$, and (S1) is nothing other than the symbol conditions (1.31) arising in the biorthogonal setting. Then the bi-frame construction would suffer from the biorthogonal restrictions. Advantageously, Theorem 4.1.3 allows for $\theta \not\equiv 1$, and this yields flexibility.

In comparison to the univariate ideas, Theorem 4.1.3 requires some particularities. The dyadic univariate setting in [CHS02, DHRS03, Chr03] does only allow for $n = 1$ in (S2) and (S2*), i.e., one directly factorizes η. We had to find a weaker concept since a convenient factorization of multivariate trigonometric polynomials is generally not possible. We found a solution by allowing for a sum of products. However, since the length of the sum determines the number of wavelets, we are still interested in small n. It turns out that if $b^{(0)} := \overline{a^{(0)}}$ is interpolatory and if it has a suitable factorization, then all additional symbols for (S2) can be constructed, and we achieve $n = 2m - 3$. Hence, in comparison to Chapter 3, Theorem 4.1.3 provides a method for reducing the number of wavelets. Finally, it should be mentioned that there still remains sufficient flexibility for the choice of $a^{(0)}$.

According to the multiplication in (S2*), some of the wavelet masks' supports are slightly larger than those of the refinable function's mask. Thus, our construction pays fewer wavelets than in Chapter 3 at the price of larger support of some wavelets. Nevertheless, we think, reducing the number of wavelets is worth the price.

Proof of Theorem 4.1.3. By applying $\eta = 1 - \theta$, we obtain

$$\begin{aligned}
&\overline{a^{(0)}(\xi+\gamma)} b^{(0)}(\xi) \theta(M^\top \xi) + \sum_{\mu=1}^{m-1+n} \overline{a^{(\mu)}(\xi+\gamma)} b^{(\mu)}(\xi) \\
&= \overline{a^{(0)}(\xi+\gamma)} b^{(0)}(\xi) \theta(M^\top \xi) + \sum_{\mu=1}^{m-1} \overline{a^{(\mu)}(\xi+\gamma)} b^{(\mu)}(\xi) \\
&\quad + \sum_{\nu=1}^{n} \overline{a^{(0)}(\xi+\gamma)} b^{(0)}(\xi) \overline{\eta^{(\nu)}(M^\top \xi)} \widetilde{\eta}^{(\nu)}(M^\top \xi) \\
&= \overline{a^{(0)}(\xi+\gamma)} b^{(0)}(\xi) \theta(M^\top \xi) + \sum_{\mu=1}^{m-1} \overline{a^{(\mu)}(\xi+\gamma)} b^{(\mu)}(\xi) \\
&\quad + \overline{a^{(0)}(\xi+\gamma)} b^{(0)}(\xi) \eta(M^\top \xi) \\
&= \overline{a^{(0)}(\xi+\gamma)} b^{(0)}(\xi) + \sum_{\mu=1}^{m-1} \overline{a^{(\mu)}(\xi+\gamma)} b^{(\mu)}(\xi) = \delta_{0,\gamma} \theta(\xi).
\end{aligned}$$

Hence, (II-c) is satisfied. Since $a^{(0)}$ and $b^{(0)}$ satisfy the sum rules of order 1 and $a^{(0)}(0) = b^{(0)}(0) = 1$, we have $\theta(0) = 1$. The number of vanishing moments is the direct consequence of the assumptions (S1*) and (S2*). $\square$

4.2 The Applicability of the General Construction Idea

4.2.1 Polyphase Conditions

In the present subsection, we express (S1) on page 70 in polyphase terms, i.e., we substitute (S1) by an equivalent condition in terms of subsymbols. Given a symbol family $\{(a^{(\mu)}, b^{(\mu)}) : \mu = 0, \dots, n\}$, recall that we denote by

$$\mathbf{a} := \left(a^{(\mu)}(\cdot + \gamma_\nu)\right)_{\substack{\nu=0,\dots,m-1\\ \mu=0,\dots,n}}, \qquad \mathbf{b} := \left(b^{(\mu)}(\cdot + \gamma_\nu)\right)_{\substack{\nu=0,\dots,m-1\\ \mu=0,\dots,n}}$$

their modulation matrices, cf. Subsection 1.1.4. Their polyphase matrices are given by

$$\mathbf{A} = \left(A^{(\mu)}_{\gamma^*_\nu}\right)_{\substack{\nu=0,\dots,m-1\\ \mu=0,\dots,n}}, \qquad \mathbf{B} = \left(B^{(\mu)}_{\gamma^*_\nu}\right)_{\substack{\nu=0,\dots,m-1\\ \mu=0,\dots,n}}$$

where $A^{(\mu)}_{\gamma^*_\nu}$ and $B^{(\mu)}_{\gamma^*_\nu}$, $\gamma^* \in \Gamma^*_M$, are the subsymbols of $a^{(\mu)}$ and $b^{(\mu)}$, respectively, see (1.39).

Then note that (S1) is equivalent to

$$\overline{\mathbf{a}}\mathbf{b}^\top = \operatorname{diag}(\theta(\cdot), \dots, \theta(\cdot + \gamma_{m-1})). \tag{4.4}$$

In order to formulate (S1) in terms of polyphases, we denote the subsymbols of θ by Θ_{γ^*}, $\gamma^* \in \Gamma^*_M$. Then the following theorem extends the condition (1.46) of the biorthogonal setting since it allows for $\theta \not\equiv 1$ and $n \geq m-1$:

Theorem 4.2.1. *Given a symbol family $\{(a^{(\mu)}, b^{(\mu)}) : \mu = 0, \dots, n\}$ and a symbol θ, then*

$$\overline{\mathbf{a}}\mathbf{b}^\top = \operatorname{diag}(\theta(\cdot), \dots, \theta(\cdot + \gamma_{m-1})) \quad \text{iff} \quad \overline{\mathbf{A}}\mathbf{B}^\top = \left(\Theta_{\gamma^*_\nu - \gamma^*_\mu}\right)_{\substack{\nu=0,\dots,m-1\\ \mu=0,\dots,m-1}}. \tag{4.5}$$

Proof. Let the matrix U be given by (1.43). Then the relations (1.44), (1.45) still hold for $n \geq m-1$. Hence, we have

$$\mathbf{a}(\xi) = \frac{1}{m} U(\xi) \mathbf{A}(M^\top \xi), \qquad \mathbf{b}(\xi) = \frac{1}{m} U(\xi) \mathbf{B}(M^\top \xi).$$

This leads to

$$\overline{\mathbf{a}(\xi)}\mathbf{b}^\top(\xi) = \frac{1}{m^2} \overline{U(\xi)}\, \overline{\mathbf{A}(M^\top \xi)} \mathbf{B}^\top(M^\top \xi) U^\top(\xi).$$

Thus, the left-hand side of (4.5) is equivalent to

$$\overline{U(\xi)}\, \overline{\mathbf{A}(M^\top \xi)} \mathbf{B}^\top(M^\top \xi) U^\top(\xi) = m^2 \operatorname{diag}(\theta(\xi), \dots, \theta(\xi + \gamma_{m-1})). \tag{4.6}$$

By applying $U\overline{U}^\top = m \cdot \mathcal{I}_m$, (4.6) is equivalent to

$$\begin{aligned}
\overline{\mathbf{A}(M^\top \xi)} \mathbf{B}^\top(M^\top \xi) &= U^\top(\xi) \operatorname{diag}(\theta(\xi), \dots, \theta(\xi + \gamma_{m-1})) \overline{U(\xi)} \\
&= \Big(\sum_{\gamma \in \Gamma_M} e^{2\pi i (\gamma^*_\mu - \gamma^*_\nu) \cdot (\xi + \gamma)} \theta(\xi + \gamma) \Big)_{\substack{\nu=0,\dots,m-1\\ \mu=0,\dots,m-1}} \\
&= \left(\Theta_{\gamma^*_\mu - \gamma^*_\nu}(M^\top \xi) \right)_{\substack{\nu=0,\dots,m-1\\ \mu=0,\dots,m-1}},
\end{aligned}$$

where the last equality is a consequence of (1.42). Since $M^\top$ is invertible, we conclude the proof. □

Remark 4.2.2. As for the equivalence of (S1) to (4.4), it can be verified that (II-c) on page 68 is equivalent to

$$\bar{\mathbf{a}}\mathbf{b}^\top = \operatorname{diag}(\theta(\cdot), \ldots, \theta(\cdot + \gamma_{m-1})),$$

where we substitute $b^{(0)}(\xi+\gamma)$ by $b^{(0)}(\xi+\gamma)\theta(M^\top\xi)$ in $\overline{\mathbf{b}}^\top$ only. By a straightforward computation, the subsymbols of $b^{(0)}(\xi)\theta(M^\top\xi)$ are given by $B^{(0)}_{\gamma^*}\theta$, for $\gamma^* \in \Gamma^*_M$. Thus, applying Theorem 4.2.1 yields the equivalence

$$(4.1) \quad \text{iff} \quad \overline{A^{(0)}_{\gamma^*_\nu}}B^{(0)}_{\gamma^*_{\nu'}}\theta + \sum_{\mu=1}^{n} \overline{A^{(\mu)}_{\gamma^*_\nu}}B^{(\mu)}_{\gamma^*_{\nu'}} = \Theta_{\gamma^*_\nu - \gamma^*_{\nu'}}, \quad \text{for } \nu, \nu' = 0, \ldots, m-1. \tag{4.7}$$

In short, we have rephrased (II-c) on page 68 in terms of subsymbols. As far as we know, (4.7) has not yet been stated elsewhere.

The right-hand side of (4.5) looks quite complicated, and the polyphase setting does not seem to provide any advantages. So far, we did not incorporate the special structure of θ. Since θ shall be given by (4.3), it is Γ_M-periodic. This simplifies the situation:

Corollary 4.2.3. *Let $\{(a^{(\mu)}, b^{(\mu)}) : \mu = 0, \ldots, n\}$ be a symbol family, and let θ be Γ_M-periodic, then* (4.5) *reduces to*

$$\bar{\mathbf{a}}\mathbf{b}^\top = \theta \cdot \mathcal{I}_m \quad \text{iff} \quad \overline{\mathbf{A}}\mathbf{B}^\top = \Theta_0 \cdot \mathcal{I}_m. \tag{4.8}$$

Proof. The collection Γ^*_M is a complete set of representatives of $\mathbb{Z}^d/M\mathbb{Z}^d$. Hence,

$$\gamma^*_\nu - \gamma^*_\mu \in M\mathbb{Z}^d \quad \text{iff} \quad \nu = \mu\,.$$

By (1.42), the Γ_M-periodicy of θ, and (1.41), we obtain

$$\begin{aligned}
\Theta_{\gamma^*_\nu - \gamma^*_\mu}(M^\top\xi) &= \sum_{\gamma\in\Gamma_M} e^{2\pi i(\gamma^*_\nu - \gamma^*_\mu)\cdot(\xi+\gamma)}\theta(\xi+\gamma) \\
&= \theta(\xi)e^{2\pi i(\gamma^*_\nu - \gamma^*_\mu)\cdot\xi} \sum_{\gamma\in\Gamma_M} e^{2\pi i(\gamma^*_\nu - \gamma^*_\mu)\cdot\gamma} \\
&= m\theta(\xi)\delta_{\nu,\mu}.
\end{aligned}$$

Thus, Theorem 4.2.1 implies (4.8). □

4.2.2 A Variant of the Matrix Completion Problem

Given $a^{(0)}$ and $b^{(0)}$, in this section, we derive additional symbols for (S1) on page 70. In order to apply the polyphase formulation of (S1) in (4.8), we need a representation of Θ_0 in terms of subsymbols of $a := a^{(0)}$ and $b := b^{(0)}$:

Lemma 4.2.4. *Given θ by* (4.3), *we have*

$$\Theta_0 = \sum_{\gamma^*\in\Gamma^*_M} \overline{A_{\gamma^*}}B_{\gamma^*}.$$

Proof. From the proof of Corollary 4.2.3, we have $\Theta_0(M^\top\xi) = m\theta(\xi)$. By applying (4.3) and (1.40), we obtain

$$\begin{aligned}\Theta_0(M^\top\xi) &= m \sum_{\gamma\in\Gamma_M} \overline{a(\xi+\gamma)}b(\xi+\gamma)\\ &= m \sum_{\gamma\in\Gamma_M} \frac{1}{m} \sum_{\gamma^*\in\Gamma_M^*} \overline{A_{\gamma^*}(M^\top\xi)e^{2\pi i\gamma^*\cdot(\xi+\gamma)}} \frac{1}{m} \sum_{\widetilde{\gamma}^*\in\Gamma_M^*} B_{\widetilde{\gamma}^*}(M^\top\xi)e^{-2\pi i\widetilde{\gamma}^*\cdot(\xi+\gamma)}\\ &= \frac{1}{m} \sum_{\gamma^*,\widetilde{\gamma}^*} \overline{A_{\gamma^*}(M^\top\xi)}B_{\widetilde{\gamma}^*}(M^\top\xi) \sum_{\gamma\in\Gamma_M} e^{-2\pi i(\widetilde{\gamma}^*-\gamma^*)\cdot(\xi+\gamma)}.\end{aligned}$$

According to (1.41), this yields

$$\begin{aligned}\Theta_0(M^\top\xi) &= \frac{1}{m} \sum_{\gamma^*,\widetilde{\gamma}^*} \overline{A_{\gamma^*}(M^\top\xi)}B_{\widetilde{\gamma}^*}(M^\top\xi)\, m\, \delta_{\widetilde{\gamma}^*,\gamma^*}\\ &= \sum_{\gamma^*\in\Gamma_M^*} \overline{A_{\gamma^*}(M^\top\xi)}B_{\gamma^*}(M^\top\xi).\end{aligned}$$

Since $M^\top$ is invertible, we conclude the proof. □

The following proposition generalizes Lemma 1.1.20. Hence, for $a^{(0)}$ interpolatory, it also generalizes the results in [JRS99]:

Proposition 4.2.5. *Given $a^{(0)}$, $b^{(0)}$, and θ by* (4.3), *we define polyphase matrices by*

$$\overline{\mathbf{A}} := \begin{pmatrix} \overline{A_0} & -B_{\gamma_1^*} & \cdots & -B_{\gamma_{m-1}^*} \\ \vdots & & & \\ \vdots & & \left(\Theta_0\mathcal{I}_{m-1} - (\overline{A_{\gamma_\nu^*}}B_{\gamma_\mu^*})_{\substack{\nu=1,\dots,m-1\\ \mu=1,\dots,m-1}}\right) & \\ \vdots & & & \\ \overline{A_{\gamma_{m-1}^*}} & & & \end{pmatrix} \tag{4.9}$$

and

$$\mathbf{B}^\top := \begin{pmatrix} B_0 & B_{\gamma_1^*} & \cdots & B_{\gamma_{m-1}^*} \\ -\overline{A_{\gamma_1^*}} & & & \\ \vdots & & \mathcal{I}_{m-1} & \\ -\overline{A_{\gamma_{m-1}^*}} & & & \end{pmatrix}. \tag{4.10}$$

If $a^{(0)}$ is interpolatory, then

$$\overline{\mathbf{A}}\mathbf{B}^\top = \Theta_0\cdot\mathcal{I}_m.$$

Proof. Since $a^{(0)}$ is interpolatory, the subsymbol A_0 equals 1, cf. [DM97]. Then the application of Lemma 4.2.4 yields

$$\left(\overline{A_0}, -B_{\gamma_1^*}, \dots, -B_{\gamma_{m-1}^*}\right)\cdot\mathbf{B}^\top = (\Theta_0, 0, \dots, 0)\,.$$

For $\nu = 1, \ldots, m-1$, we obtain with $A_0 \equiv 1$

$$\left(\overline{A_{\gamma_\nu^*}}, -\overline{A_{\gamma_\nu^*}} B_{\gamma_1^*}, \ldots, \Theta_0 - \overline{A_{\gamma_\nu^*}} B_{\gamma_\nu^*}, -\overline{A_{\gamma_\nu^*}} B_{\gamma_{\nu+1}^*}, \ldots\right) \cdot \begin{pmatrix} B_0 \\ -\overline{A_{\gamma_1^*}} \\ \vdots \\ -\overline{A_{\gamma_{m-1}^*}} \end{pmatrix}$$
$$= \overline{A_{\gamma_\nu^*}} \Big(\sum_{\mu=0}^{m-1} \overline{A_{\gamma_\mu^*}} B_{\gamma_\mu^*} - \Theta_0 \Big) = 0.$$

Finally, we calculate

$$\left(\overline{A_{\gamma_\nu^*}}, -\overline{A_{\gamma_\nu^*}} B_{\gamma_1^*}, \ldots, \Theta_0 - \overline{A_{\gamma_\nu^*}} B_{\gamma_\nu^*}, -\overline{A_{\gamma_\nu^*}} B_{\gamma_{\nu+1}^*}, \ldots\right) \cdot \begin{pmatrix} \overline{B_{\gamma_1^*}} & \cdots & \overline{B_{\gamma_{m-1}^*}} \\ 1 & & 0 \\ & \ddots & \\ 0 & & 1 \end{pmatrix}$$
$$= \left(0, \ldots, 0, \overline{A_{\gamma_\nu^*}} B_{\gamma_\nu^*} + \Theta_0 - \overline{A_{\gamma_\nu^*}} B_{\gamma_\nu^*}, 0 \ldots, 0\right)$$
$$= (0, \ldots, 0, \Theta_0, 0, \ldots, 0).$$

□

Due to (1.40), the polyphase matrices (4.9), (4.10) implicitly provide symbols $a^{(1)}$, ..., $a^{(m-1)}$ and $b^{(1)}$, ..., $b^{(m-1)}$, which satisfy (S1). They are specified in the following corollary:

Corollary 4.2.6 (Step 1). *Using the notation of Theorem 4.1.3, let the symbols $a^{(0)}$ and $b^{(0)}$ satisfy the sum rules of order $2s$ and $a^{(0)}(0) = b^{(0)}(0) = 1$. Additionally, let $a^{(0)}$ be interpolatory. For $\mu = 1, \ldots, m-1$, we define*

$$a^{(\mu)}(\xi) = e^{-2\pi i \gamma_\mu^* \cdot \xi} \sum_{\gamma \in \Gamma_M \setminus \{0\}} \left(a^{(0)}(\xi + \gamma) - a^{(0)}(\xi) e^{-2\pi i \gamma_\mu^* \cdot \gamma} \right) \overline{b^{(0)}(\xi + \gamma)}, \tag{4.11}$$

$$b^{(\mu)}(\xi) = \frac{1}{m} e^{-2\pi i \gamma_\mu^* \cdot \xi} \sum_{\gamma \in \Gamma_M \setminus \{0\}} \left(1 - e^{-2\pi i \gamma_\mu^* \cdot \gamma} \right) \overline{a^{(0)}(\xi + \gamma)}. \tag{4.12}$$

Then (S1) in Theorem 4.1.3 is satisfied, and (S1) holds for $s_1 = 2s$.*

Proof. Let $a^{(1)}$, ..., $a^{(m-1)}$ and $b^{(1)}$, ..., $b^{(m-1)}$ be the symbols associated to the polyphase matrices (4.9), (4.10). In the following, we verify that they coincide with (4.11) and (4.12), respectively.

By applying (1.42) and (1.40), a direct computation yields

$$a^{(\mu)}(\xi) = e^{-2\pi i \gamma_\mu^* \cdot \xi} \overline{\theta(\xi)} - a^{(0)}(\xi) \sum_{\gamma \in \Gamma_M} e^{-2\pi i \gamma_\mu^* \cdot (\xi + \gamma)} \overline{b^{(0)}(\xi + \gamma)},$$

$$b^{(\mu)}(\xi) = \frac{1}{m} \Big(e^{-2\pi i \gamma_\mu^* \cdot \xi} - \sum_{\gamma \in \Gamma_M} e^{-2\pi i \gamma_\mu^* \cdot (\xi + \gamma)} \overline{a^{(0)}(\xi + \gamma)} \Big).$$

The definition of θ in (4.3) leads to

$$\begin{aligned} a^{(\mu)}(\xi) &= e^{-2\pi i \gamma_\mu^* \cdot \xi}\Big(\sum_{\gamma \in \Gamma_M} \left(a^{(0)} \overline{b^{(0)}}\right)(\xi+\gamma) - a^{(0)}(\xi) \sum_{\gamma \in \Gamma_M} e^{-2\pi i \gamma_\mu^* \cdot \gamma} \overline{b^{(0)}(\xi+\gamma)}\Big) \\ &= e^{-2\pi i \gamma_\mu^* \cdot \xi}\Big(\sum_{\gamma \in \Gamma_M \setminus \{0\}} \left(a^{(0)} \overline{b^{(0)}}\right)(\xi+\gamma) - a^{(0)}(\xi) \sum_{\gamma \in \Gamma_M \setminus \{0\}} e^{-2\pi i \gamma_\mu^* \cdot \gamma} \overline{b^{(0)}(\xi+\gamma)}\Big) \\ &= e^{-2\pi i \gamma_\mu^* \cdot \xi} \sum_{\gamma \in \Gamma_M \setminus \{0\}} \left(a^{(0)}(\xi+\gamma) - a^{(0)}(\xi) e^{-2\pi i \gamma_\mu^* \cdot \gamma}\right) \overline{b^{(0)}(\xi+\gamma)}. \end{aligned}$$

Since the symbol $\overline{a^{(0)}}$ is interpolatory, we obtain

$$\begin{aligned} b^{(\mu)}(\xi) &= \frac{1}{m} e^{-2\pi i \gamma_\mu^* \cdot \xi}\Big(\sum_{\gamma \in \Gamma_M} \overline{a^{(0)}(\xi+\gamma)} - \sum_{\gamma \in \Gamma_M} e^{-2\pi i \gamma_\mu^* \cdot \gamma} \overline{a^{(0)}(\xi+\gamma)}\Big) \\ &= \frac{1}{m} e^{-2\pi i \gamma_\mu^* \cdot \xi}\Big(\sum_{\gamma \in \Gamma_M \setminus \{0\}} \overline{a^{(0)}(\xi+\gamma)} - \sum_{\gamma \in \Gamma_M \setminus \{0\}} e^{-2\pi i \gamma_\mu^* \cdot \gamma} \overline{a^{(0)}(\xi+\gamma)}\Big) \\ &= \frac{1}{m} e^{-2\pi i \gamma_\mu^* \cdot \xi} \sum_{\gamma \in \Gamma_M \setminus \{0\}} \left(1 - e^{-2\pi i \gamma_\mu^* \cdot \gamma}\right) \overline{a^{(0)}(\xi+\gamma)}. \end{aligned}$$

So far, $\mathbf{A}$ and $\mathbf{B}$ in Proposition 4.2.5 are the associated polyphase matrices of the symbol family $\{(a^{(\mu)}, b^{(\mu)}) : \mu = 0, \dots, m-1\}$ given by (4.11) and (4.12). By applying Proposition 4.2.5 and Corollary 4.2.3, the associated modulation matrices $\mathbf{a}$ and $\mathbf{b}$ satisfy (S1).

According to the representations (4.11) and (4.12), the sum rules of $a^{(0)}$ and $b^{(0)}$ imply vanishing moments of the wavelet symbols. □

Remark 4.2.7. Although $b^{(0)}$ is not necessarily dual to $a^{(0)}$, Corollary 4.2.6 provides the same symbols as Lemma 1.1.20. Moreover, provided that $m = 2$, a direct computation yields that the symbols

$$a^{(1)}(\xi) = e^{-2\pi i \gamma^* \cdot \xi} \overline{b^{(0)}(\xi+\gamma)}, \qquad b^{(1)}(\xi) = e^{-2\pi i \gamma^* \cdot \xi} \overline{a^{(0)}(\xi+\gamma)} \tag{4.13}$$

solve (S1). In this situation, we do not have to assume that $a^{(0)}$ is interpolatory. Hence, even Example 1.1.19 provides a solution to (S1).

Proposition 4.2.5 as well as the proof of Corollary 4.2.6 is a simple combinatorial approach. Certainly, Corollary 4.2.6 can be directly proven without any polyphase decomposition, see Appendix A.2 for details.

Next, we derive an alternative solution of (S1), which also addresses noninterpolatory symbols. The results in [RS92] yield the following lemma with respect to dyadic dilation. It even provides the coincidence of primal and dual wavelet symbols.

Lemma 4.2.8. *Let $M = 2\mathcal{I}_d$. Given a real-valued symbol $a^{(0)}$, let there exist a function $\vartheta : \{0,1\}^d \to \{0,1\}^d$ with $\vartheta(0) = 0$ such that, for all $\gamma_1^*, \gamma_2^* \in \{0,1\}^d$, $\gamma_1^* \neq \gamma_2^*$, the usual inner product on $\mathbb{R}^d$*

$$\left(\vartheta(\gamma_1^*) + \vartheta(\gamma_2^*)\right) \cdot \left(\gamma_1^* + \gamma_2^*\right) \tag{4.14}$$

is odd. Then the choice

$$a^{(\gamma^*)}(\xi) := e^{-2\pi i \vartheta(\gamma^*) \cdot \xi} a^{(0)}(\xi + \tfrac{1}{2}\gamma^*), \qquad \text{for } \gamma^* \in \{0,1\}^d,$$

yields a family $\{a^{(\gamma^*)} : \gamma^* \in \{0,1\}^d\}$, whose modulation matrix $\mathbf{a}$ satisfies $\overline{\mathbf{a}}\mathbf{a}^\top = \theta \cdot \mathcal{I}_{2^d}$, where θ is given by (4.3) with $b^{(0)} = a^{(0)}$.

In the univariate setting, a possible choice for ϑ in Lemma 4.2.8 is $\vartheta(0) = 0$ and $\vartheta(1) = 1$. A bivariate choice is given by $\vartheta(0) = 0$ and

$$(1,0) \mapsto (1,1), \quad (0,1) \mapsto (0,1), \quad (1,1) \mapsto (1,0).$$

In case $d = 3$, the mapping $\vartheta(0) = 0$ with

$$\begin{array}{lll} (1,0,0) \mapsto (1,1,0), & (0,1,0) \mapsto (0,1,1), & (1,1,0) \mapsto (1,0,0), \\ (0,0,1) \mapsto (1,0,1), & (1,0,1) \mapsto (0,0,1), & (0,1,1) \mapsto (0,1,0), \\ (1,1,1) \mapsto (1,1,1), & & \end{array}$$

satisfies (4.14), see [RS92]. However, according to [RS91], there does not exist a mapping $\vartheta : \{0,1\}^d \to \{0,1\}^d$ such that (4.14) holds for $d > 3$. Thus, Lemma 4.2.8 only provides a solution of (S1) for $d = 1, 2$, and 3.

Let us summarize the results of the present subsection. We can solve (S1) provided that $a^{(0)}$ is interpolatory, see Corollary 4.2.6. For $m = 2$, we also find a solution for noninterpolatory $a^{(0)}$, cf. Remark 4.2.7. In case of dyadic dilation and real-valued $a^{(0)}$, Lemma 4.2.8 yields symbols for (S1) provided that the dimension is less than 4.

4.2.3 Splitting into a Sum of Products

In [LS], Lai and Stöckler construct bivariate compactly supported tight wavelet frames from box spline refinable functions. Their construction is different from Theorem 4.1.3, but there is a similarity to (S2). For box spline symbols $b^{(0)} = a^{(0)}$ and η given by (4.3), they decompose η into a sum of squares, i.e., they find symbols $\eta^{(\nu)}$, $\nu = 1, \ldots, n$, such that

$$\eta = \sum_{\nu=1}^{n} \left|\eta^{(\nu)}\right|^2 .$$

Thus, their approach is more restrictive than our decomposition into a sum of products. In fact, their construction really suffers from this restriction since their resulting compactly supported wavelets have only one vanishing moment.

We use our add-on in flexibility to obtain a high number of vanishing moments. Interpolatory symbols play a key role:

Lemma 4.2.9. *Under the assumptions of Corollary* 4.2.6 *and the notation of Theorem* 4.1.3, *let also* $b^{(0)}$ *be interpolatory. Then,* η *has* $2s$ *vanishing moments.*

Proof. Since both symbols $a^{(0)}, b^{(0)}$ are interpolatory, we obtain

$$\begin{aligned} 1 &= \sum_{\gamma \in \Gamma_M} \overline{a^{(0)}(\xi+\gamma)} \sum_{\tilde{\gamma} \in \Gamma_M} b^{(0)}(\xi+\tilde{\gamma}) \\ &= \sum_{\gamma \in \Gamma_M} \overline{a^{(0)}(\xi+\gamma)} b^{(0)}(\xi+\gamma) + \sum_{\gamma \neq \tilde{\gamma}} \overline{a^{(0)}(\xi+\gamma)} b^{(0)}(\xi+\tilde{\gamma}). \end{aligned}$$

This leads to

$$\eta(\xi) = \sum_{\gamma \neq \tilde{\gamma}} \overline{a^{(0)}(\xi+\gamma)} b^{(0)}(\xi+\tilde{\gamma}). \tag{4.15}$$

Applying the sum rules, we obtain that η has $2s$ vanishing moments. □

In (4.15), we already have a sum of products but it does not lead to vanishing moments because $a^{(0)}(0) = b^{(0)}(0) = 1$. If $a^{(0)}$ has a suitable factorization, then we obtain $\eta^{(\nu)}$ and $\widetilde{\eta}^{(\nu)}$, $\nu = 1, \dots, n$, with a high number of vanishing moments:

Proposition 4.2.10 (Step 2). *Under the assumptions of Corollary 4.2.6 and the notation of Theorem 4.1.3, let $b^{(0)} = \overline{a^{(0)}}$. Moreover, let there exist symbols $c^{(0)}$ and $d^{(0)}$ satisfying the sum rules of order s such that $a^{(0)} = \overline{c^{(0)} d^{(0)}}$. For $\mu = 1, \dots, m-1$ and $\nu = 1, \dots, m-2$, we define*

$$\begin{aligned} \eta^{(\mu)}(\xi) &:= \overline{c^{(0)}(\xi)} d^{(0)}(\xi + \gamma_\mu), & \widetilde{\eta}^{(\mu)}(\xi) &:= \overline{2 \overline{d^{(0)}(\xi)} c^{(0)}(\xi + \gamma_\mu)}, \\ \eta^{(m-1+\nu)}(\xi) &:= a^{(0)}(\xi + \gamma_\nu), & \widetilde{\eta}^{(m-1+\nu)}(\xi) &:= \overline{2 \textstyle\sum_{j=\nu+1}^{m-1} a^{(0)}(\xi + \gamma_j)}. \end{aligned}$$

Then, (S2) and (S2) in Theorem 4.1.3 are satisfied with $n = 2m - 3$ and $s_2 = s$.*

Proof. Since $b^{(0)} = \overline{a^{(0)}}$, the assumptions of Lemma 4.2.9 are satisfied. Thus, we can apply (4.15):

$$\begin{aligned} \eta(\xi) &= \sum_{\gamma \neq \tilde{\gamma}} (c^{(0)} \overline{d^{(0)}})(\xi + \gamma)(\overline{d^{(0)}} c^{(0)})(\xi + \tilde{\gamma}) \\ &= 2 \sum_{0 \leq i < j \leq m-1} c^{(0)}(\xi + \gamma_i) \overline{d^{(0)}(\xi + \gamma_j)}\, \overline{d^{(0)}(\xi + \gamma_i) c^{(0)}(\xi + \gamma_j)} \\ &= \sum_{\mu=1}^{m-1} c^{(0)}(\xi) \overline{d^{(0)}(\xi + \gamma_\mu) 2 \overline{d^{(0)}(\xi)} c^{(0)}(\xi + \gamma_\mu)} \\ &\quad + \sum_{\nu=1}^{m-2} c^{(0)}(\xi + \gamma_\nu) \overline{d^{(0)}(\xi + \gamma_\nu) 2 \sum_{j=\nu+1}^{m-1} c^{(0)}(\xi + \gamma_j) \overline{c^{(0)}(\xi + \gamma_j)}} \\ &= \sum_{\nu=1}^{n} \overline{\eta^{(\nu)}(\xi) \widetilde{\eta}^{(\nu)}}(\xi). \end{aligned}$$

The sum rules for $c^{(0)}$ and $d^{(0)}$ imply the vanishing moments in (S2*). □

Steps 1 and 2 provide all components required for Theorem 4.1.3. It yields $3m - 4$ wavelets, and they have at least s vanishing moments. Thus, in comparison to Chapter 3, we reduced the number of wavelets from $m^2 - 1$ to linear dependency on m. For the desirable case $m = 2$, we reduce three wavelets in Chapter 3 to two wavelets in the present chapter.

Remark 4.2.11. The assumption about the factorization of the symbol $a^{(0)}$ is not as restrictive as it seems. Most of the well-known interpolatory symbols are obtained by some iteration process. Thus, the symbols $c^{(0)}$ and $d^{(0)}$ are already known, see for example [Der99, JRS99] as well as Lemma 3.2.3, Section 3.3, and the following Section 4.3 in the present thesis.

4.3 Examples of Optimal Wavelet Bi-Frames

In the sequel, we present wavelet bi-frames constructed from the refinable functions addressed in Section 3.3. The first example provides arbitrarily smooth wavelet bi-frames

N	1	2	3	4
mask $a^{(0)}$	$\diamondsuit_3$	$\diamondsuit_7$	$\diamondsuit_{11}$	$\diamondsuit_{15}$
s_∞	2	3.5	4.7	≥ 5.8
sum rules	4	8	12	16
vm	2	4	6	8
card(wavelets)	2	2	2	2

optimality	
sum rules	yes
smoothness	yes
card(wavelets)	yes
approx. order	yes
symmetry	yes

Table 4.1: Optimal multivariate bi-frame Checkerboard $(N)_\mathrm{R}$

in arbitrary dimensions with only two wavelets. They satisfy all optimality conditions introduced in Section 2.3.

Example 4.3.1 (Checkerboard). We recall the notation for the construction of the Checkerboard bi-frames in Subsection 3.3.1. Let

$$\begin{aligned}\breve{a}(z) &:= \left(\frac{1+z}{2}\right)^N \cdot \left(\frac{1+1/z}{2}\right)^N,\\ \breve{b}(z) &:= \left(\frac{1+z}{2}\right)^N \cdot \left(\frac{1+1/z}{2}\right)^N \cdot P_{2N}\left(1-\left(\frac{1+z}{2}\right)\cdot\left(\frac{1+1/z}{2}\right)\right)\end{aligned}$$

denote univariate symbols. We define multivariate symbols by

$$a^{(0)}(\xi_1,\ldots,\xi_d) := \breve{a}(\xi_1), \qquad b^{(0)}(\xi_1,\ldots,\xi_d) := \breve{b}(\xi_1),$$

and let M be given by (2.27), (2.28). Since the univariate symbol $\breve{a}\breve{b}$ is interpolatory and $\Gamma_M = \{0, (\frac{1}{2},\ldots,\frac{1}{2})^\top\}$, see (2.17), the multivariate symbol $a^{(0)}b^{(0)}$ is interpolatory with respect to M, and Step 1 is applicable. Since $a^{(0)}b^{(0)}$ trivially inherits a suitable factorization, Step 2 is also applicable. Thus, we obtain a wavelet bi-frame from the same refinable function as the bi-frame Checkerboard (N) in Section 3.3. Since it only requires two wavelets, we call the bi-frame *Checkerboard* $(N)_\mathrm{R}$, where R stands for "reduced number of wavelets". According to the discussion in Subsection 4.1.1, two wavelets are regarded as the optimal number of wavelets.

It should be mentioned that the wavelet symbols still inherit the symmetry of $a^{(0)}b^{(0)}$. For increasing N, we obtain arbitrarily smooth bi-frames in arbitrary dimensions. They are optimal with respect to all criteria considered in Subsection 2.3.5, see Table 4.1 for details.

Next, we address the quincunx dilation matrix. We obtain wavelet bi-frames with only two wavelets from the refinable function of the Laplace (N_1-N_2) bi-frame:

Example 4.3.2 (Laplace). First, we recall the construction of the underlying refinable function of Laplace (N_1-N_2). Given the quincunx dilation matrix

$$M = \begin{pmatrix} 1 & -1 \\ 1 & 1 \end{pmatrix},$$

let

$$\widetilde{c}(z_1,z_2) := \frac{1}{2}\left(1+\frac{1}{4}z_1+\frac{1}{4}z_1^{-1}+\frac{1}{4}z_2+\frac{1}{4}z_2^{-1}\right)$$

N	1	2
mask $a^{(0)}$	$\Diamond_3$	$\Diamond_7$
s_∞	1.3	2.5
sum rules	4	8
vm	2	4
card(wavelets)	2	2

optimality	
sum rules	yes
smoothness	–
card(wavelets)	yes
approx. order	yes
symmetry	yes

Table 4.2: Laplace $(N\text{-}N)_{\mathrm{R}}$

be the Laplace symbol. We choose

$$\begin{aligned} c^{(0)} &:= \widetilde{c}^{N_1}, \\ d^{(0)} &:= \widetilde{c}^{N_2} P_{N_1+N_2}(1-\widetilde{c}). \end{aligned}$$

Then, $a^{(0)} = c^{(0)}d^{(0)}$ is interpolatory, and it trivially inherits a suitable factorization. Hence, both Step 1 and Step 2 are applicable. They provide a wavelet bi-frame with only two wavelets, which we call *Laplace* $(N_1\text{-}N_2)_{\mathrm{R}}$.

As for Laplace $(N_1\text{-}N_2)$, the wavelets inherit the full symmetries of the underlying refinable function. We also have optimality with respect to sum rules, number of wavelets, and approximation order, see Table 4.2. Figure 4.2 shows the refinable function and the wavelets of Laplace $(1\text{-}1)_{\mathrm{R}}$.

In the following example, we address fundamental box splines obtained in [RS97a].

Example 4.3.3 (Box Spline)**.** Let $M = 2\mathcal{I}_2$ and, for $N = 1, 2, 3$, and 4, let

$$c^{(0)}(z) := \left(\frac{1+z_1}{2}\right)^N \cdot \left(\frac{1+z_2}{2}\right)^N \cdot \left(\frac{1+z_1z_2}{2}\right)^N \cdot \left(\frac{1}{z_1z_2}\right)^N$$

be the symbol of the three-direction box spline with equal multiplicities N. In [RS97a], a symbol $\widetilde{d}$ was constructed such that

$$a^{(0)}(z) := \left(\frac{1+z_1}{2}\right)^{2N} \cdot \left(\frac{1+z_2}{2}\right)^{2N} \cdot \left(\frac{1+z_1z_2}{2}\right)^{2N} \cdot \left(\frac{1}{z_1z_2}\right)^{2N} \cdot \widetilde{d}(z) \tag{4.16}$$

is interpolatory and nonnegative. Hence, we have the factorization

$$a^{(0)} = c^{(0)}d^{(0)}, \tag{4.17}$$

where $d^{(0)} = c^{(0)}\widetilde{d}$. Notice that $c^{(0)}$ and $d^{(0)}$ are real-valued and nonnegative. It turns out that the mask of $a^{(0)}$ is $\mathcal{G}$-symmetric with

$$\mathcal{G} = \left\{\pm\mathcal{I}_2, \pm\begin{pmatrix}0 & 1\\ 1 & 0\end{pmatrix}, \pm\begin{pmatrix}1 & 0\\ 1 & -1\end{pmatrix}, \pm\begin{pmatrix}0 & 1\\ -1 & 1\end{pmatrix}, \pm\begin{pmatrix}1 & -1\\ 1 & 0\end{pmatrix}, \pm\begin{pmatrix}-1 & 1\\ 0 & 1\end{pmatrix}\right\}.$$

Applying Lemma 4.2.8 yields that

$$\begin{aligned} a^{(1)}(z_1,z_2) &:= z_1z_2a^{(0)}(-z_1,z_2), \\ a^{(2)}(z_1,z_2) &:= z_2a^{(0)}(z_1,-z_2), \\ a^{(3)}(z_1,z_2) &:= z_1a^{(0)}(-z_1,-z_2). \end{aligned} \tag{4.18}$$

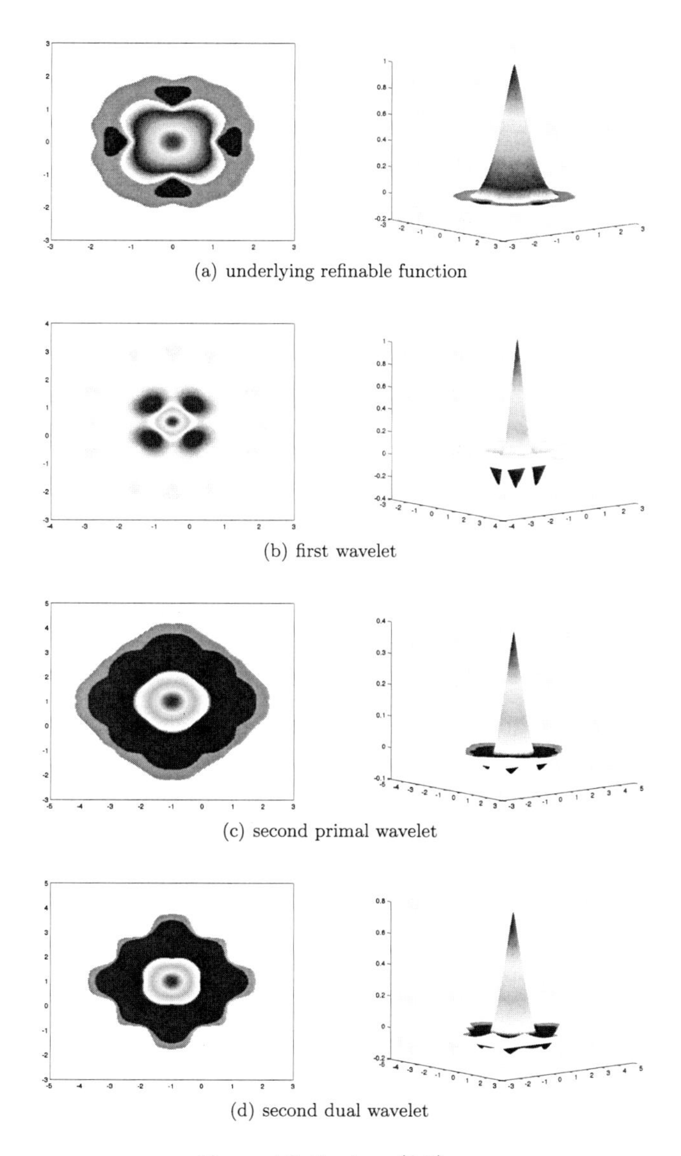

(a) underlying refinable function

(b) first wavelet

(c) second primal wavelet

(d) second dual wavelet

Figure 4.2: Laplace $(1\text{-}1)_{\mathrm{R}}$

N	1	2	3	4
mask $a^{(0)}$	$\square_3$	$\square_7$	$\square_{11}$	$\square_{15}$
s_∞	2	3.5	4.7	≥ 5.8
sum rules	4	8	12	16
vm	2	4	6	8
card(wavelets)	8	8	8	8

optimality	
sum rules	yes
smoothness	yes
card(wavelets)	–
approx. order	yes
symmetry	–

Table 4.3: Box Spline $(N)_{\mathrm{R}}$ of Example 4.3.3

satisfies

$$\bar{\mathbf{a}}\mathbf{a}^\top = \theta \cdot \mathcal{I}_4,$$

where θ is given by (4.3) with $b^{(0)} = a^{(0)}$. Then let $\mathbf{b} := \mathbf{a}$. Hence, our first three primal and dual wavelets coincide. By using (4.17), Step 2 is applicable. We obtain a bi-frame with 8 wavelets, which we call *Box Spline $(N)_{\mathrm{R}}$*. The wavelet masks are $\mathcal{G}_i$-symmetric for different subgroups $\mathcal{G}_i \subset \mathcal{G}$:

$$\mathcal{G}_1 := \left\{ \pm\mathcal{I}_2, \pm \begin{pmatrix} 0 & 1 \\ 1 & 0 \end{pmatrix} \right\}, \qquad a^{(3)}, a^{(6)}, b^{(6)}, b^{(8)} \text{ are } \mathcal{G}_1\text{-symmetric.}$$

$$\mathcal{G}_2 := \left\{ \pm\mathcal{I}_2, \pm \begin{pmatrix} -1 & 1 \\ 0 & 1 \end{pmatrix} \right\}, \qquad a^{(2)}, a^{(5)}, b^{(5)}, a^{(8)} \text{ are } \mathcal{G}_2\text{-symmetric.}$$

$$\mathcal{G}_3 := \left\{ \pm\mathcal{I}_2, \pm \begin{pmatrix} 1 & 0 \\ 1 & -1 \end{pmatrix} \right\}, \qquad a^{(1)}, a^{(4)}, b^{(4)}, a^{(7)}, b^{(7)} \text{ are } \mathcal{G}_3\text{-symmetric.}$$

Thus, the bi-frame does not have optimal symmetry. Nevertheless, we have optimality with respect to sum rules and approximation order. Moreover, we obtain the same estimates of the $L_\infty(\mathbb{R}^2)$-critical exponent s_∞ as for the univariate reference function in Theorem 2.3.6, which is optimal, see Table 4.3 for properties and optimality results.

Let us compare Box Spline $(N)_{\mathrm{R}}$ with the construction in Subsection 3.3.3. There, we derived a wavelet bi-frame Box Spline (1) from the same refinable function as Box Spline $(1)_{\mathrm{R}}$. However, Box Spline (1) requires 15 wavelets, and some of them are only symmetric with respect to $\mathcal{G}_4 = \{\pm\mathcal{I}_2\}$. Hence, the approach of the present example provides more symmetry, fewer wavelets, and it is applicable to smoother box splines.

Chapter 5

The Characterization of Function Spaces

Function spaces are related to many fields of applied mathematics. For instance, operator equations, partial differential equations, and many kinds of variational problems require the consideration of smoothness classes, cf. [DD97, DDU02, DDD04]. In order to derive fast practical algorithms for the solution of such problems, a discretization is required. Thus, one looks for certain basis-like systems, which characterize the smoothness class. This means that each f in the space has a series expansion, that there is a predefined rule for the choice of its coefficients, and that the original smoothness norm is equivalent to a sequence norm of coefficients obtained from this rule.

Many smoothness spaces allow for nicely structured discretizations. For instance, Modulation spaces are characterized by Gabor systems, wavelets constitute the above mentioned basis-like systems in Besov spaces, and so-called Gabor wavelets provide a characterization of Alpha-Modulation spaces, see the textbooks [FS03, Grö01] and references therein.

In the aforementioned problems, Besov spaces are the arising smootheness classes. The characterization of Besov spaces by dyadic orthonormal wavelet bases was derived by DeVore, Jawerth, and Popov in the early nineties, cf. [DJP92]. It turns out that the range of smoothness parameters of the Besov spaces is restricted by the smoothness of the wavelets and their vanishing moments. Lindemann extended the characterization to pairs of biorthogonal wavelet bases with general isotropic scalings, see [Lin05]. Recently, Borup, Gribonval, and Nielsen characterized Besov spaces by wavelet bi-frames, cf. [BGN04]. However, their results are restricted to dyadic dilation. In the present chapter, we extend the results to general isotropic dilation matrices.

To point out the difficulties of the extension, we shall explain the main idea of the bi-frame approach in [BGN04]. Initially, one chooses a dyadic orthonormal basis characterizing the Besov space. Recall that the characterization requires sufficient smoothness, and one can choose a tensor product of Meyer wavelets or of sufficiently smooth Daubechies wavelets, cf. [Dau92]. Then one applies a certain localization technique, i.e., the bi-frame is localized to the dyadic orthonormal wavelet basis such that the orthonormal characterization carries over to the wavelet bi-frame, see Subsection 5.3.1 for details. Hence, the orthonormal basis plays the role of a reference system. In order to address general isotropic scalings, there arise two problems. First, we have to extend the localization technique from dyadic to isotropic dilation. Second, for many isotropic dilation matrices, it is not clear whether there exist smooth compactly supported orthogonal wavelets. Hence, we need another reference system.

We proceed as follows: first, we introduce Besov spaces in detail. Then we recall their characterization by pairs of biorthogonal wavelet bases with general isotropic scalings.

Since, for most of the known dilation matrices, there exist smooth compactly supported biorthogonal wavelets, see for instance [Der99, JRS99], they constitute promising substitutes for the orthogonal wavelet basis. Next, we generalize the localization technique regarding isotropic dilation and biorthogonal reference systems. Then the biorthogonal characterization carries over to the wavelet bi-frame. Finally, we apply the results to our bi-frames constructed in the previous Chapters 3 and 4.

5.1 Besov Spaces

The present section is dedicated to the introduction of Besov spaces on $\mathbb{R}^d$. They are divided into homogeneous and nonhomogeneous spaces. Nevertheless, for the range of smoothness parameters, which are required for our characterization by wavelets, they essentially describe the same functions class. First, we recall the nonhomogeneous spaces.

5.1.1 Nonhomogeneous Besov Spaces

Nonhomogeneous Besov spaces on $\mathbb{R}^d$ can be defined in two different ways. In approximation theory, one generally describes the spaces by means of differences since it allows for a simple and direct extension to arbitrary domains. The harmonic analysis approach applies Fourier based methods to define Besov spaces. Fortunately, both approaches are equivalent for the range of parameters we are addressing. First, following [Tri83, RS96], we recall the Fourier approach.

Let $\Phi(\mathbb{R}^d)$ be the collection of all $\phi = (\phi_j)_{j\in\mathbb{N}_0}$ contained in the Schwartz space $\mathcal{S}(\mathbb{R}^d)$, such that the following holds:

(i) there exist positive constants A, B, C with

$$\begin{aligned}
&\operatorname{supp}(\phi_0) \subset \{x \in \mathbb{R}^d : \|x\| \leq A\},\\
&\operatorname{supp}(\phi_j) \subset \{x \in \mathbb{R}^d : B2^{j-1} \leq \|x\| \leq C2^{j+1}\}, \quad \text{for } j = 1,2,3,\dots,
\end{aligned}$$

(ii) for every $\alpha \in \mathbb{N}_0^d$,

$$\sup_{x\in\mathbb{R}^d} \sup_{j\in\mathbb{N}_0} 2^{j|\alpha|} \left|\partial^\alpha \phi_j(x)\right| < \infty,$$

(iii) and, for all $x \in \mathbb{R}^d$,

$$\sum_{j=0}^{\infty} \phi_j(x) = 1.$$

Let $s \in \mathbb{R}$, $0 < p, q \leq \infty$, and $\phi \in \Phi$, then

$$B^s_{p,q}(\mathbb{R}^d) := \left\{ f \in \mathcal{S}'(\mathbb{R}^d) : \|f\|^{\phi}_{B^s_{p,q}} < \infty \right\}$$

is called the *nonhomogeneous Besov space*, where

$$\|f\|^{\phi}_{B^s_{p,q}} := \left\| \left(2^{js} \left\| \mathcal{F}^{-1}(\phi_j \mathcal{F} f) \right\|_{L_p} \right)_{j\in\mathbb{N}_0} \right\|_{\ell_q}. \tag{5.1}$$

Here, $\mathcal{F}$ denotes the Fourier transform on the space of tempered distributions $\mathcal{S}'(\mathbb{R}^d)$, cf. Appendix A.1. It turns out that the definition of $B^s_{p,q}(\mathbb{R}^d)$ does not depend on the choice of ϕ, and different choices yield equivalent expressions, see [RS96, Tri83]. We do not distinguish between them, and we write $\|f\|_{B^s_{p,q}}$ instead of $\|f\|^{\phi}_{B^s_{p,q}}$. The space $B^s_{p,q}(\mathbb{R}^d)$ is complete, hence, for $1 \leq p < \infty$, $1 \leq q \leq \infty$, it is a Banach space. For $0 < p < 1$ or $0 < q < 1$, the expression in (5.1), is only a *quasi-norm*, i.e., the triangular inequality holds up to a constant factor, see for instance [Tri83], and we have a quasi Banach space.

Many known function spaces, such as Sobolev, Hölder, and Lipschitz spaces, are covered by the scale of Besov spaces. For instance, we have with equivalent norms

$$\begin{aligned} W^s(L_2(\mathbb{R}^d)) &= B^s_{2,2}(\mathbb{R}^d), \quad 0 < s, \\ W^s(L_p(\mathbb{R}^d)) &= B^s_{p,p}(\mathbb{R}^d), \quad 0 < s \notin \mathbb{N},\ 1 \leq p < \infty, \\ C^s(\mathbb{R}^d) &= B^s_{\infty,\infty}(\mathbb{R}^d), \quad 0 < s \notin \mathbb{N}, \\ \operatorname{Lip}\left(s, L_p(\mathbb{R}^d)\right) &= B^s_{p,\infty}(\mathbb{R}^d), \quad 0 < s \notin \mathbb{N},\ 1 \leq p \leq \infty, \end{aligned}$$

see [Coh03, DL93, RS96, Tri83]. The index q is of minor interest, and Besov spaces are mainly determined by s and p since, for $0 < \varepsilon$,

$$B^{s+\varepsilon}_{p,\infty}(\mathbb{R}^d) \hookrightarrow B^s_{p,q}(\mathbb{R}^d), \tag{5.2}$$

$$B^s_{p,q}(\mathbb{R}^d) \hookrightarrow B^s_{p,q+\varepsilon}(\mathbb{R}^d) \hookrightarrow B^s_{p,\infty}(\mathbb{R}^d), \tag{5.3}$$

see for instance Section 2.3 in [Tri83] or Chapter 2 in [RS96]. For fixed $1 < p < \infty$, we apply the short-hand notation

$$B^s := B^s_{\tau,\tau}(\mathbb{R}^d), \quad \tfrac{1}{\tau} = \tfrac{s}{d} + \tfrac{1}{p},\ 0 < s. \tag{5.4}$$

These spaces arise in the context of nonlinear approximation as described in Chapter 6 of the present thesis. Then given $1 < p < \infty$ and $0 < \varepsilon$, the following embeddings hold,

$$B^{s+\varepsilon} \hookrightarrow B^s, \tag{5.5}$$

$$B^s \hookrightarrow L_p(\mathbb{R}^d), \tag{5.6}$$

cf. Chapter 2 in [RS96]. Figure 5.1 provides a visualization.

Nonhomogeneous Besov Spaces by Means of Differences

In approximation theory, it is often more convenient to work with a definition of Besov spaces by means of differences since it allows for a simple and unified definition on arbitrary domains. Nevertheless, we only require $\mathbb{R}^d$. For $f : \mathbb{R}^d \to \mathbb{C}$ and $h \in \mathbb{R}^d$, the difference of order l with $l \in \mathbb{N}$ is recursively given by

$$\Delta^1_h f := f(\cdot + h) - f(\cdot), \quad \Delta^l_h := \Delta^1_h \Delta^{l-1}_h.$$

Then, for $f \in L_p(\mathbb{R}^d)$, $0 < p \leq \infty$, and $t \geq 0$, we define the *modulus of smoothness* of order l by

$$\omega_l(f,t)_{L_p} := \sup_{|h| \leq t} \left\| \Delta^l_h f \right\|_{L_p}. \tag{5.7}$$

Note that the difference of order 1 has already been used in the context of Lipschitz spaces, and the modulus of smoothness of order 1 equals the modulus of continuity, see Appendix A.1.

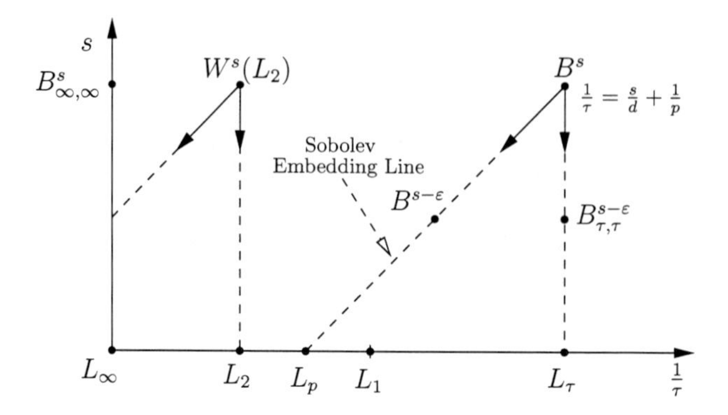

Figure 5.1: Embeddings of Besov Spaces.

Definition 5.1.1. For $0 < p, q \leq \infty$, $0 < s < \infty$, the *Besov space* $B^s_q(L_p(\mathbb{R}^d))$ is the collection of all $f \in L_p(\mathbb{R}^d)$ such that the *Besov semi-norm*

$$|f|_{B^s_q(L_p)} := \begin{cases} \left(\int_0^\infty \left(t^{-s}\omega_l(f,t)_{L_p}\right)^q \frac{dt}{t}\right)^{\frac{1}{q}}, & \text{for } 0 < q < \infty, \\ \sup_{0<t<\infty}(t^{-s}\omega_l(f,t)_{L_p}), & \text{for } q = \infty, \end{cases} \tag{5.8}$$

is finite, where l is an integer such that $l > s$.

It should be mentioned that, for $0 < p < 1$ or $0 < q < 1$, the Besov semi-norm is only a quasi-semi-norm. Different choices of l provide equivalent semi-norms, and the Besov space $B^s_q(L_p(\mathbb{R}^d))$ is equipped with the *Besov norm*

$$\|f\|_{B^s_q(L_p(\mathbb{R}^d))} := \|f\|_{L_p} + |f|_{B^s_q(L_p)}, \tag{5.9}$$

cf. Chapter 2 in [DL93]. For $0 < p < 1$ or $0 < q < 1$, expression (5.9) is only a quasi-norm, but in the following, we suppress these distinctions. It is often more convenient to work with the following equivalent discretization of the original Besov semi-norm, i.e.,

$$|f|_{B^s_q(L_p)} \sim \left\| \left(2^{sj}\omega_l(f, 2^{-j})_{L_p}\right)_{j\in\mathbb{Z}} \right\|_{\ell_q}. \tag{5.10}$$

The index set $\mathbb{Z}$ can also be substituted by $\mathbb{N}_0$, cf. Chapter 2 in [DL93], and according to the results of Section 4.3 in [Lin05], the number 2 in (5.10) can be replaced by any fixed number greater than 1.

Let us explore the spaces $B^s_q(L_p(\mathbb{R}^d))$ in more detail. The parameter s measures smoothness in $L_p(\mathbb{R}^d)$. In this sense, they are closely related to the fractional Sobolev spaces $W^s(L_p(\mathbb{R}^d))$, cf. Appendix A.1. For $0 < \varepsilon$, we have the embeddings as in (5.2), (5.3), and (5.6),

$$B^{s+\varepsilon}_\infty(L_p(\mathbb{R}^d)) \hookrightarrow B^s_q(L_p(\mathbb{R}^d)), \tag{5.11}$$

$$B^s_q(L_p(\mathbb{R}^d)) \hookrightarrow B^s_{q+\varepsilon}(L_p(\mathbb{R}^d)) \hookrightarrow B^s_\infty(L_p(\mathbb{R}^d)), \tag{5.12}$$

$$B^s_\tau(L_\tau(\mathbb{R}^d)) \hookrightarrow L_p, \quad 1 < p < \infty, \ \tfrac{1}{\tau} = \tfrac{s}{d} + \tfrac{1}{p}, \tag{5.13}$$

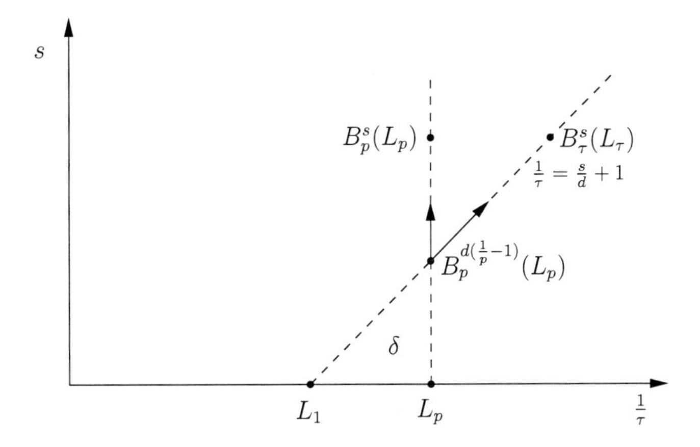

Figure 5.2: Range of Besov Spaces with $B^s_{p,q}(\mathbb{R}^d) = B^s_q(L_p(\mathbb{R}^d))$.

cf. Chapters 2 and 12 in [DL93] or Chapters 3 and 4 in [Coh03].

Let us finally determine the range of parameters, for which the Fourier bases approach and the method by means of differences yield the same spaces. For $0 < s$, $1 \le p \le \infty$, and $0 < q \le \infty$, we have the identity

$$B^s_q(L_p(\mathbb{R}^d)) = B^s_{p,q}(\mathbb{R}^d), \tag{5.14}$$

with equivalent norms, see Section 2.5 in [Tri83] or Chapter 2 in [RS96]. Hence, the new definition of Besov spaces coincides with our old one. An extension of the identity (5.14) to the range $0 < p < 1$ needs some fine distinctions. For $0 < p < 1$, $0 < s$, and $\frac{s}{d} + 1 < \frac{1}{p}$, the Dirac distribution δ is contained in $B^s_{p,q}(\mathbb{R}^d)$, see Section 2.5.3 in [Tri83]. Then (5.14) does not hold since $\delta \notin L_p(\mathbb{R}^d)$. For the reverse inequality $\frac{s}{d} + 1 > \frac{1}{p}$, the spaces coincide, see Section 2.3 in [RS96] or Section 2.5.12 in [Tri83] for the following result:

Proposition 5.1.2. *Given $0 < s$, we have with equivalent norms*

$$B^s_q(L_p(\mathbb{R}^d)) = B^s_{p,q}(\mathbb{R}^d), \quad \textit{for } 0 < p, q \le \infty \textit{ and } \tfrac{s}{d} + 1 > \tfrac{1}{p}. \tag{5.15}$$

Note that (5.15) includes the spaces B^s in (5.4), i.e., for $1 < p < \infty$,

$$B^s = B^s_\tau(L_\tau(\mathbb{R}^d)), \quad \tfrac{1}{\tau} = \tfrac{s}{d} + \tfrac{1}{p}.$$

See Figure 5.2 for a visualization of the above discussion.

5.1.2 Homogeneous Besov Spaces

Similar to their nonhomogeneous counterparts, homogeneous Besov spaces can be described by some Fourier based approach and by means of differences. The characterization by wavelets as well as Chapters 6 and 7 address a range of smoothness parameters, where both approaches yield identical spaces. Moreover, for this range, homogeneous and nonhomogeneous spaces essentially describe the same functions.

First, we recall the Fourier based approach, and we modify $\Phi(\mathbb{R}^d)$ in Subsection 5.1.1. In short, we replace $\mathbb{N}_0$ by $\mathbb{Z}$. Let $\dot{\Phi}(\mathbb{R}^d)$ be the collection of all $\phi = (\phi_j)_{j\in\mathbb{Z}}$ contained in the Schwartz space $\mathcal{S}(\mathbb{R}^d)$, such that the following holds:

(i) there exist positive constants B, C with

$$\operatorname{supp}(\phi_j) \subset \{x \in \mathbb{R}^d : B2^{j-1} \leq \|x\| \leq C2^{j+1}\}, \quad \text{for } j \in \mathbb{Z},$$

(ii) for every $\alpha \in \mathbb{N}_0^d$,

$$\sup_{x\in\mathbb{R}^d} \sup_{j\in\mathbb{Z}} 2^{j|\alpha|} \left|\partial^\alpha \phi_j(x)\right| < \infty,$$

(iii) and, for all $x \in \mathbb{R}^d$,

$$\sum_{j\in\mathbb{Z}} \phi_j(x) = 1.$$

Similar to the nonhomogeneous Besov norm, for $f \in \mathcal{S}'(\mathbb{R}^d)$, we consider

$$\left\| \left(2^{js} \left\| \mathcal{F}^{-1}\left(\phi_j \mathcal{F} f\right)\right\|_{L_p} \right)_{j\in\mathbb{Z}} \right\|_{\ell_q}. \tag{5.16}$$

According to [RS96, Tri83, Tri92], f is a polynomial iff (5.16) equals zero. Hence, it is natural to consider the quotient $\mathcal{S}'(\mathbb{R}^d)/\Pi(\mathbb{R}^d)$, where $\Pi(\mathbb{R}^d)$ denotes the collection of all d-dimensional polynomials. This space is essentially a dual space, which we shall specify: given

$$\mathcal{Z}(\mathbb{R}^d) := \left\{ f \in \mathcal{S}(\mathbb{R}^d) : \left(\partial^\alpha \mathcal{F} f\right)(0) = 0, \text{for all } \alpha \in \mathbb{N}_0^d \right\}$$

equipped with the induced topology of the Schwartz space, its topological dual $\mathcal{Z}'(\mathbb{R}^d)$ can be identified with $\mathcal{S}'(\mathbb{R}^d)/\Pi(\mathbb{R}^d)$, see Chapter 5 in [Tri83] for details. Let $\phi \in \dot{\Phi}$ be fixed, then, for $s \in \mathbb{R}$ and $0 < p, q \leq \infty$,

$$\dot{B}^s_{p,q}(\mathbb{R}^d) := \left\{ f \in \mathcal{Z}'(\mathbb{R}^d) : \|f\|^{\phi}_{\dot{B}^s_{p,q}} < \infty \right\}$$

is called the *homogeneous Besov space*, where $\|f\|^{\phi}_{\dot{B}^s_{p,q}}$ is given by (5.16). Similar to its nonhomogeneous counterpart, different choices of ϕ provide equivalent expressions. Hence, we may write $\|f\|_{\dot{B}^s_{p,q}}$ instead of $\|f\|^{\phi}_{\dot{B}^s_{p,q}}$. Then $\dot{B}^s_{p,q}(\mathbb{R}^d)$ is complete, but for $0 < q < 1$ or $0 < p < 1$, the expression $\|\cdot\|_{\dot{B}^s_{p,q}}$ is only a quasi-norm, see Section 2.6 in [RS96] or Section 5.1 in [Tri83]. Finally, the estimate

$$\|f(\lambda\cdot)\|_{\dot{B}^s_{p,q}} \lesssim \lambda^{s-\frac{d}{p}} \|f\|_{\dot{B}^s_{p,q}},$$

for all $\lambda > 0$ and $f \in \dot{B}^s_{p,q}(\mathbb{R}^d)$, justifies the term "homogeneous", cf. Section 2.6 in [RS96] or Section 5.1 in [Tri83].

Comparing the definitions of homogeneous and nonhomogeneous Besov space, one may expect some close relation between them. In fact, for a large range of parameters, they essentially describe the same spaces, see for instance Section 2.6 in [RS96] for the following result:

Proposition 5.1.3. *For $0 < s$, $0 < p, q \leq \infty$ with $\frac{s}{d} + 1 > \frac{1}{p}$, we have*

$$\|f\|_{B^s_{p,q}} \sim \left(\|f\|_{\dot{B}^s_{p,q}} + \|f\|_{L_p}\right).$$

Hence, for the range of parameters in Proposition 5.1.3, $B^s_{p,q}(\mathbb{R}^d)$ equals $\dot{B}^s_{p,q}(\mathbb{R}^d) \cap L_p(\mathbb{R}^d)$ equipped with the usual norm of the intersection space.

Homogeneous Besov Spaces by means of Differences

Let $\mathscr{H}$ be a quasi-normed complete, translation invariant subspace of $\mathcal{S}'(\mathbb{R}^d)$ or of complex-valued functions on $\mathbb{R}^d$. Then the l-th modulus of smoothness in (5.7) can be extended by

$$\omega_l(f,t)_{\mathscr{H}} := \sup_{|h|\leq t} \left\|\Delta^l_h f\right\|_{\mathscr{H}}.$$

Following Kyriazis in [Kyr96], for $0 < s < \infty$ and $0 < q \leq \infty$, denote $\dot{B}^s_q(\mathscr{H})$ the collection of all $f \in \mathscr{H}$ such that

$$\|f\|_{\mathscr{H}} := \begin{cases} \left(\int_0^\infty (t^{-s}\omega_l(f,t)_{\mathscr{H}})^q \frac{dt}{t}\right)^{\frac{1}{q}}, & \text{for } 0 < q < \infty, \\ \sup_{0<t<\infty}(t^{-s}\omega_l(f,t)_{\mathscr{H}}), & \text{for } q = \infty, \end{cases}$$

is finite, where $l > s$. Different choices of l provide equivalent expressions. Hence, for $\mathscr{H} = L_p(\mathbb{R}^d)$, the space $\dot{B}^s_q(L_p(\mathbb{R}^d))$ essentially contains the same functions as its nonhomogeneous counterpart $B^s_q(L_p(\mathbb{R}^d))$ in Definition 5.1.1. However, $\dot{B}^s_q(L_p(\mathbb{R}^d))$ is equipped with the Besov semi-norm, i.e., $\|f\|_{\dot{B}^s_q(L_p)} = |f|_{B^s_q(L_p)}$. In [Kyr96], Kyriazis explores the equivalence of the homogeneous Besov spaces $\dot{B}^s_q(H_p(\mathbb{R}^d))$, $\dot{B}^s_q(L_p(\mathbb{R}^d))$, and $\dot{B}^s_{p,q}(\mathbb{R}^d)$, cf. Appendix A.1 for the Hardy space $H_p(\mathbb{R}^d)$:

Theorem 5.1.4. *For $0 < s < \infty$, $0 < q \leq \infty$, and $0 < p < \infty$, we have*

$$\dot{B}^s_q(H_p(\mathbb{R}^d)) = \dot{B}^s_{p,q}(\mathbb{R}^d) \tag{5.17}$$

with equivalent norms. If we additionally restrict the range of parameters to $\frac{s}{d} + 1 > \frac{1}{p}$, then (5.17) also holds with respect to $\dot{B}^s_q(L_p(\mathbb{R}^d))$.

Next, we introduce the homogeneous counterpart of the notation (5.4): given $1 < p < \infty$, let

$$\dot{B}^s := \dot{B}^s_\tau(L_\tau(\mathbb{R}^d)), \quad \frac{1}{\tau} = \frac{s}{d} + \frac{1}{p}. \tag{5.18}$$

Since $1 < p < \infty$, we have $\frac{s}{d} + 1 > \frac{1}{\tau}$. This provides

$$\dot{B}^s = \dot{B}^s_{\tau,\tau}(\mathbb{R}^d), \quad \frac{1}{\tau} = \frac{s}{d} + \frac{1}{p}. \tag{5.19}$$

By applying (5.19), the results in [Jaw77] yield the continuous embeddings

$$\dot{B}^{s+\varepsilon} \hookrightarrow \dot{B}^s, \quad \text{for all } \varepsilon > 0,$$

and according to [DJP92], we have

$$\dot{B}^s \hookrightarrow L_p(\mathbb{R}^d), \tag{5.20}$$

which constitutes a stronger version of (5.6).

5.2 The Characterization by Means of Biorthogononal Wavelets

Homogeneous Besov spaces are characterized in [DJP92] by dyadic orthonormal wavelet bases, see also the survey article [DeV98]. In [Lin05], Lindemann extends the characterization to biorthogonal wavelets with general isotropic scalings. However, he considers nonhomogeneous Besov spaces. Then contrary to [DJP92], the wavelet coefficients are only involved on the scales $j \in \mathbb{N}_0$. The negative scales are covered by the coefficients of the underlying refinable function. In the sequel, we characterize the homogeneous Besov space by pairs of biorthogonal wavelet bases with general isotropic scalings. Then the underlying refinable function is not involved, and one incorporates wavelet coefficients on the entire range of scales $j \in \mathbb{Z}$.

The homogeneous Besov space $\dot{B}^s$ is embedded in $L_p(\mathbb{R}^d)$, see (5.20). In order to derive a convenient framework for further considerations, we extend the Definition 1.1.9 of a multiresolution analysis in $L_2(\mathbb{R}^d)$ to $L_p(\mathbb{R}^d)$. According to (M-4), this requires an extension of stability. We say a function $f \in L_p(\mathbb{R}^d)$ is *ℓ_p-stable* if, for all $(c_k)_{k\in\mathbb{Z}^d} \in \ell_p(\mathbb{Z}^d)$, the series $\sum_{k\in\mathbb{Z}^d} c_k\varphi_{0,k}$ converges in $L_p(\mathbb{R}^d)$ and

$$\left\| \sum_{k\in\mathbb{Z}^d} c_k\varphi_{0,k} \right\|_{L_p} \sim \left\| (c_k)_{k\in\mathbb{Z}^d} \right\|_{\ell_p}.$$

For compactly supported functions, ℓ_p- and ℓ_2-stability are equivalent, see [JM90] for the following theorem.

Theorem 5.2.1. *Given $1 \leq p \leq \infty$ and a compactly supported function $f \in L_2(\mathbb{R}^d) \cap L_p(\mathbb{R}^d)$, then f is ℓ_p-stable iff it is ℓ_2-stable.*

Next, we extend the multiresolution analysis framework to $L_p(\mathbb{R}^d)$:

Definition 5.2.2. Given $0 < p < \infty$, an increasing sequence $(V_j)_{j\in\mathbb{Z}}$ of closed subspaces in $L_p(\mathbb{R}^d)$ is called a *multiresolution analysis in $L_p(\mathbb{R}^d)$* if the following holds:

(M-1) $f \in V_j$ iff $f(M^{-j}\cdot) \in V_0$,

(M-2) $\bigcup_{j\in\mathbb{Z}} V_j$ is dense in $L_p(\mathbb{R}^d)$,

(M-3) $\bigcap_{j\in\mathbb{Z}} V_j = \{0\}$,

(M-4) there exists an ℓ_p-stable $\varphi \in V_0$ such that V_0 is the closed linear span of its integer shifts.

The function φ in (M-4) is called the *generator* of the multiresolution analysis.

Let φ be a compactly supported generator of a multiresolution analysis in $L_2(\mathbb{R}^d)$. Given $1 \leq p \leq \infty$, if φ is also contained in $L_p(\mathbb{R}^d)$, then it is ℓ_p-stable, cf. Theorem 5.2.1. Hence, by defining

$$V_j := \operatorname{clos}_{L_p}\left(\operatorname{span}\left\{\varphi_{j,k} : k \in \mathbb{Z}^d\right\}\right), \quad j \in \mathbb{Z}, \tag{5.21}$$

φ constitutes a candidate for a generator of a multiresolution analysis in $L_p(\mathbb{R}^d)$. However, in order to obtain sufficient conditions, we still require some preparation. Given a nonempty open subset $A \subset \mathbb{R}^d$, we say a compactly supported distribution φ has *linearly independent integer shifts on* A if

$$\sum_{k \in \mathbb{Z}^d} c_k \varphi(\cdot - k) = 0 \quad \text{on } A,$$

implies

$$c_k \varphi(\cdot - k) = 0 \quad \text{on } A, \text{ for all } k \in \mathbb{Z}^d.$$

Moreover, recall that for an isotropic dilation matrix M, we denote the modulus of its eigenvalues by ρ. The following proposition is one of the fundamental results in [Lin05]:

Proposition 5.2.3. *Suppose that M is isotropic. Given $0 < p < \infty$ and $0 < q \leq \infty$, let $\varphi \in L_p(\mathbb{R}^d)$ be a compactly supported refinable function. Additionally, let the integer shifts of φ be linearly independent on a bounded open cube in $\mathbb{R}^d$, and let $(V_j)_{j \in \mathbb{Z}}$ be as in* (5.21)*. Then the following holds:*

(a) *If φ reproduces polynomials up to order l, then*

$$\operatorname{dist}(f, V_j)_{L_p} \lesssim \omega_l(f, \rho^{-j})_{L_p}, \quad \text{for all } f \in L_p(\mathbb{R}^d) \text{ and } j \in \mathbb{Z}. \tag{5.22}$$

(b) *If $\varphi \in W^s(L_\infty(\mathbb{R}^d))$, $s \in \mathbb{N}$, then for $s < l \in \mathbb{N}$,*

$$\omega_l(f, t)_{L_p} \lesssim \min\{1, t\rho^j\}^s \|f\|_{L_p}, \quad \text{for all } f \in V_j \text{ and } j \in \mathbb{Z},\ t \geq 0. \tag{5.23}$$

Let φ satisfy the assumptions of Proposition 5.2.3. One easily verifies that linear independence on some cube implies global linear independence in (1.36). Then according to [JM90], φ is ℓ_p-stable. Thus, the spaces V_j in (5.21) satisfy (M-1) and (M-4). Due to the results in [CT97], the inequalities (5.22) and (5.23) in Proposition 5.2.3 imply (M-2) and (M-3), respectively. Hence, φ is a generator of a multiresolution analysis in $L_p(\mathbb{R}^d)$.

Remark 5.2.4. According to [Coh03, Lin05], the assumption about linear independence over some cube is always satisfied if there exists a second compactly supported refinable function $\widetilde{\varphi}$, which has biorthogonal integer shifts with respect to φ. Hence, Proposition 5.2.3 is applicable in the setting of biorthogonal wavelet bases. Moreover, then φ is stable, which implies $\widehat{\varphi}(0) \neq 0$, see for instance [Lin05]. Since φ is compactly supported, the assumption in part (b) yields $\varphi \in W^s(L_1(\mathbb{R}^d))$. Thus, by applying Theorem 1.2.6, φ reproduces polynomials up to order l with $s < l \leq s+1$. Then the assumption of part (b) implies the requirements of part (a).

Proposition 5.2.3 also provides some kind of characterization of Besov spaces by a multiresolution analysis, see [DJP92] for the dyadic case of the following theorem and [Lin05] for a different version addressing the nonhomogeneous Besov norm:

Theorem 5.2.5. *Under the assumptions of Proposition 5.2.3, let φ be contained in $W^s(L_\infty(\mathbb{R}^d))$, $s \in \mathbb{N}$, and suppose $\widehat{\varphi}(0) \neq 0$. Then, for $0 < \alpha < s$ and for all $f \in L_p(\mathbb{R}^d)$,*

$$\|f\|_{\dot{B}^\alpha_q(L_p)} \sim \left\| \left(\rho^{\alpha j} \operatorname{dist}(f, V_j)_{L_p}\right)_{j \in \mathbb{Z}} \right\|_{\ell_q}. \tag{5.24}$$

In comparison to the dyadic setting, the main difficulty of the proof of Theorem 5.2.6 with respect to isotropic dilation is the verification of Proposition 5.2.3. If the proposition is once established, then we may follow DeVore, Jawerth, and Popov in [DJP92], and we derive the theorem. For the sake of completeness, we present a detailed proof in Appendix A.2.

Biorthogonal wavelet bases yield complementary spaces for the underlying multiresolution analysis in $L_2(\mathbb{R}^d)$, see Subsection 1.1.3. According to the results in [CT97], these decompositions also hold in $L_p(\mathbb{R}^d)$, and then applying Theorem 5.2.5 provides the characterization of Besov spaces in terms of biorthogonal wavelets. However, before we can explicitly state the result in the following theorem, we still require some preparation. Let

$$\psi_{j,k}^{(\mu),p}(x) := m^{\frac{j}{p}} \psi^{(\mu)}(M^j x - k), \quad \text{for } j \in \mathbb{Z},\ k \in \mathbb{Z}^d,\ x \in \mathbb{R}^d,$$

denote the $L_p(\mathbb{R}^d)$-normalization of $\psi_{j,k}^{(\mu)}$, and let us use the short-hand notation

$$\psi_\lambda^p = \psi_{\mu,j,k}^p := \psi_{j,k}^{(\mu),p},$$

where $\lambda = (\mu, j, k)$ and

$$\Lambda := \{1, \ldots, m-1\} \times \mathbb{Z} \times \mathbb{Z}^d.$$

We do so for the dual wavelets as well. Then according to [CT97], Theorem 5.2.5 implies the following characterization of homogeneous Besov spaces:

Theorem 5.2.6. *Let M be an isotropic dilation matrix, and let $1 < p < \infty$, $\frac{1}{p} + \frac{1}{p'} = 1$. Suppose that*

$$X(\{\psi^{(1)}, \ldots, \psi^{(m-1)}\}), \quad X(\{\widetilde{\psi}^{(1)}, \ldots, \widetilde{\psi}^{(m-1)}\})$$

are a pair of compactly supported biorthogonal wavelet bases, whose underlying refinable functions φ and $\widetilde{\varphi}$ are contained in $L_p(\mathbb{R}^d)$ and $L_{p'}(\mathbb{R}^d)$, respectively. Let φ be also contained in $W^s(L_\infty(\mathbb{R}^d))$, $s \in \mathbb{N}$. Then, for all $0 < \alpha < s$ and $f \in \dot{B}^\alpha$, the series expansion

$$f = \sum_{\lambda \in \Lambda} \left\langle f, \widetilde{\psi}_\lambda^{p'} \right\rangle \psi_\lambda^p \tag{5.25}$$

holds in $L_p(\mathbb{R}^d)$, and

$$\|f\|_{\dot{B}^\alpha} \sim \left\| \left(\left\langle f, \widetilde{\psi}_\lambda^{p'} \right\rangle \right)_{\lambda \in \Lambda} \right\|_{\ell_\tau}, \tag{5.26}$$

where $\frac{1}{\tau} = \frac{\alpha}{d} + \frac{1}{p}$.

It should be mentioned that the inner products in (5.25) and (5.26) make sense as the duality mappings between $L_p(\mathbb{R}^d)$ and $L_{p'}(\mathbb{R}^d)$ since $\dot{B}^\alpha$ is a subset of $L_p(\mathbb{R}^d)$, see (5.20). Moreover, according to the norm equivalences (5.26), the partial sums of the series (5.25) constitute a Cauchy sequence in $\dot{B}^\alpha$. Since this space is complete, the series also converges in $\dot{B}^\alpha$.

5.3 The Characterization by Means of Wavelet Bi-Frames

In Theorem 5.2.5, we established an equivalence between the homogeneous Besov norm and a weigthed sequence norm involving the multiresolution analysis. Since biorthogonal wavelets yield complementary decompositions of the underlying multiresolution analysis, this approach provides the characterization in terms of a pair of biorthogonal wavelet bases in Theorem 5.2.6, cf. [CT97]. Wavelet bi-frames do generally not provide such a complementary decomposition. Thus, the characterization of Besov spaces by this weaker concept requires a different tool.

In a series of papers, Gröchenig et al. consider localized frames, i.e., frames, whose Gramian matrices have certain decay outside the diagonal, see [Grö03, Grö04, GC04, GF05]. In some sense, we follow these ideas. We also address Gramian type matrices, and we estimate the decay of their entries outside the diagonal. However, we apply localization to two different frames, i.e., we consider their mixed Gramian matrices. Finally, we establish that the mixed Gramian matrix of bi-frame wavelets and biorthogonal wavelets constitutes a bounded operator on certain sequence spaces. Then by applying some results about wavelet bi-frame expansions in $L_p(\mathbb{R}^d)$, the biorthogonal characterization carries over to the bi-frame.

5.3.1 A Localization by the Mixed Gramian

Let $\{f_\kappa : \kappa \in \mathcal{K}\}$, $\{g_{\kappa'} : \kappa' \in \mathcal{K}'\}$ be two Bessel sequences in a Hilbert space $\mathcal{H}$. Then their synthesis operators F and G are bounded, see Subsection 2.1.1. Thus, G^*F is a bounded operator on $\ell_2(\mathcal{K})$. It coincides with the mixed Gramian matrix operator

$$(c_\kappa)_{\kappa\in\mathcal{K}} \mapsto \left(\sum_{\kappa\in\mathcal{K}} \langle f_\kappa, g_{\kappa'} \rangle \, c_\kappa \right)_{\kappa'\in\mathcal{K}'}.$$

The following theorem shows that, for wavelet systems, the mixed Gramian is bounded on a large scale of ℓ_τ-spaces. It is our main result of this section, and it extends the dyadic results in [BGN04] to general isotropic scalings. Note that we do not assume strong differentiability as they do in [BGN04]. We only require weak differentiability.

Theorem 5.3.1. *Let M be isotropic and $s, s' \in \mathbb{N}$. For $\mu \in E := \{1, \dots, n\}$ and $\mu' \in E' := \{1, \dots, n'\}$, let compactly supported functions*

$$f^{(\mu)} \in W^s(L_\infty(\mathbb{R}^d)) \quad \text{and} \quad g^{(\mu')} \in W^{s'}(L_\infty(\mathbb{R}^d))$$

have s' and s vanishing moments, respectively. Given $1 \le p < \infty$ and $1 = \frac{1}{p} + \frac{1}{p'}$, we consider the matrix operator

$$T : (c_\lambda)_{\lambda\in\Lambda} \mapsto \left(\sum_{\lambda\in\Lambda} \left\langle f_\lambda^p, g_{\lambda'}^{p'} \right\rangle c_\lambda \right)_{\lambda'\in\Lambda'},$$

where $\Lambda = E \times \mathbb{Z} \times \mathbb{Z}^d$ and $\Lambda' = E' \times \mathbb{Z} \times \mathbb{Z}^d$. Then $T : \ell_\tau(\Lambda) \to \ell_\tau(\Lambda')$ is bounded for any τ in the range

$$p\left(\frac{s'}{d} + 1\right) > \tau > \begin{cases} \left(\frac{s}{d} + \frac{1}{p}\right)^{-1}, & \text{for } \frac{s}{d} + \frac{1}{p} \ge 1, \\ p\left(1 - \frac{s}{d}\right), & \text{for } \frac{s}{d} + \frac{1}{p} \le 1. \end{cases}$$

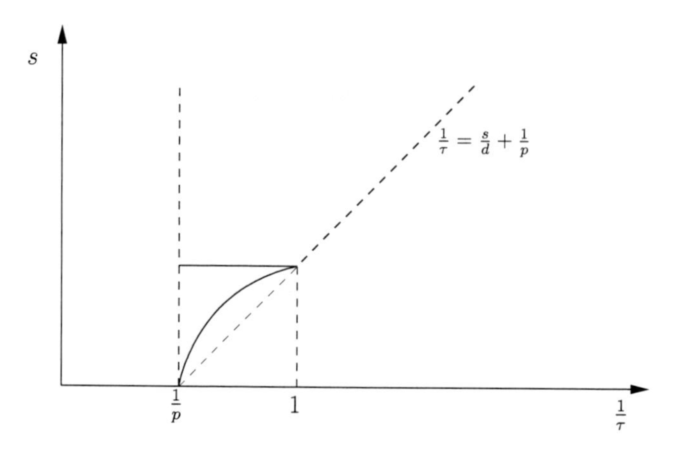

Figure 5.3: Range of τ

For the application of Theorem 5.3.1 in Subsection 5.3.3, the upper bound of τ is of minor interest since we only require $p \geq \tau$. However, the lower bound is critical, and it yields a restriction. Unfortunately, it can not be improved in general, see [BGN04] for a counterexample. A visualization is given in Figure 5.3.

The proof of Theorem 5.3.1 keeps us busy for the remainder of the present subsection. One of the two fundamental ingredients is the following lemma. It extends the dyadic Lemma 8.10 in [MC97] also allowing for isotropic dilation matrices.

Lemma 5.3.2. *Let M be isotropic, and let $d < \delta$. For $j \in \mathbb{Z}$, consider the matrix operator T_j given by*

$$(d_k)_{k\in\mathbb{Z}^d} \mapsto \begin{cases} \left(\sum_{k\in\mathbb{Z}^d} \left(1 + \|k - M^{-j}k'\|\right)^{-\delta} d_k\right)_{k'\in\mathbb{Z}^d}, & \text{for } j > 0, \\ \left(\sum_{k\in\mathbb{Z}^d} \left(1 + \|M^j k - k'\|\right)^{-\delta} d_k\right)_{k'\in\mathbb{Z}^d}, & \text{for } j \leq 0. \end{cases}$$

Then, T_j is bounded on $\ell_\tau(\mathbb{Z}^d)$, for any $1 \leq \tau \leq \infty$, and its operator norm satisfies

$$\|T_j\|_{\ell_\tau \to \ell_\tau} \lesssim \begin{cases} m^{\frac{j}{\tau}}, & j > 0, \\ m^{-\frac{j}{\tau'}}, & j \leq 0, \end{cases}$$

where $\frac{1}{\tau} + \frac{1}{\tau'} = 1$.

Proof. First, we address $j \leq 0$, and we consider $\tau = 1$ and $\tau = \infty$. For $1 < \tau < \infty$, we apply the Riesz-Thorin Interpolation Theorem, see Theorem A.3.1 in the Appendix.

Let us choose $\tau = 1$. In order to derive

$$\|T_j\|_{\ell_1 \to \ell_1} \lesssim 1, \tag{5.27}$$

we split $M^j k$ into the sum $l + r$ with $\|r\|_\infty < 1$, where $\|r\|_\infty$ denotes the maximum norm on $\mathbb{R}^d$. This yields

$$\sum_{k'\in\mathbb{Z}^d}(1+\|M^jk-k'\|)^{-\delta}\lesssim\sum_{k'\in\mathbb{Z}^d}(1+\|l+r-k'\|_\infty)^{-\delta}$$
$$=\sum_{k'\in\mathbb{Z}^d}(1+\|r-k'\|_\infty)^{-\delta}.$$

Applying the reverse triangle inequality $|\|r\|_\infty-\|k'\|_\infty|\le\|r-k'\|_\infty$ provides

$$\sum_{k'\in\mathbb{Z}^d}(1+\|M^jk-k'\|)^{-\delta}\lesssim\sum_{k'\in\mathbb{Z}^d}\left(1+\left|\|r\|_\infty-\|k'\|_\infty\right|\right)^{-\delta}.$$

Since $\|r\|_\infty<1$ and $d<\delta$, we obtain

$$\sum_{k'\in\mathbb{Z}^d}(1+\|M^jk-k'\|)^{-\delta}\lesssim\sum_{k'\in\mathbb{Z}^d\backslash\{0\}}(\|k'\|_\infty)^{-\delta}+1\lesssim 1.$$

For $(d_k)_{k\in\mathbb{Z}^d}\in\ell_1(\mathbb{Z}^d)$, this yields

$$\left\|T_j\left((d_k)_{k\in\mathbb{Z}^d}\right)\right\|_{\ell_1}=\sum_{k'\in\mathbb{Z}^d}\left|\sum_{k\in\mathbb{Z}^d}(1+\|M^jk-k'\|)^{-\delta}d_k\right|$$
$$\le\sum_{k\in\mathbb{Z}^d}\sum_{k'\in\mathbb{Z}^d}(1+\|M^jk-k'\|)^{-\delta}|d_k|$$
$$\lesssim\left\|(d_k)_{k\in\mathbb{Z}^d}\right\|_{\ell_1}.$$

Thus, (5.27) holds.

Now, let us address $\tau=\infty$. In the following, we verify

$$\|T_j\|_{\ell_\infty\to\ell_\infty}\lesssim m^{-j}. \tag{5.28}$$

This requires the introduction of a special norm: for isotropic dilation matrices M, there exists a norm $\|\cdot\|_M$ on $\mathbb{R}^d$ such that

$$\|Mx\|_M=\rho\|x\|_M,\quad\text{for all } x\in\mathbb{R}^d, \tag{5.29}$$

where ρ is the modulus of the eigenvalues of M, cf. [Jia98]. Since all norms on $\mathbb{R}^d$ are equivalent, this leads to

$$\sum_{k\in\mathbb{Z}^d}m^j(1+\|M^jk-k'\|)^{-\delta}\lesssim\sum_{k\in\mathbb{Z}^d}m^j(1+\|M^jk-k'\|_M)^{-\delta}$$
$$=\sum_{k\in\mathbb{Z}^d}m^j(1+\|M^j(k-M^{-j}k')\|_M)^{-\delta}.$$

Due to $j\le 0$, we have $M^{-j}k'\in\mathbb{Z}^d$. This provides with $m^j=\rho^{jd}$

$$\sum_{k\in\mathbb{Z}^d}m^j(1+\|M^jk-k'\|)^{-\delta}=\sum_{k\in\mathbb{Z}^d}m^j(1+\|M^jk\|_M)^{-\delta}$$
$$\lesssim\sum_{k\in\mathbb{Z}^d}\rho^{jd}(1+\|\rho^jk\|)^{-\delta}.$$

Since the last term is a Riemann sum of the integrable function $x \mapsto (1 + \|x\|)^{-\delta}$, we obtain

$$\sum_{k \in \mathbb{Z}^d} m^j (1 + \|M^j k - k'\|)^{-\delta} \lesssim 1.$$

For $(d_k)_{k \in \mathbb{Z}^d} \in \ell_\infty(\mathbb{Z}^d)$, the Cauchy-Schwartz inequality and the last estimate imply

$$\begin{aligned}
\left\| T_j \left((d_k)_{k \in \mathbb{Z}^d} \right) \right\|_{\ell_\infty \to \ell_\infty} &= \sup_{k' \in \mathbb{Z}^d} \left| \sum_{k \in \mathbb{Z}^d} (1 + \|M^j k - k'\|)^{-\delta} d_k \right| \\
&\leq m^{-j} \sup_{k' \in \mathbb{Z}^d} \sum_{k \in \mathbb{Z}^d} m^j (1 + \|M^j k - k'\|)^{-\delta} |d_k| \\
&= m^{-j} \sup_{k' \in \mathbb{Z}^d} \left\| \left(m^j (1 + \|M^j k - k'\|)^{-\delta} d_k \right)_{k \in \mathbb{Z}^d} \right\|_{\ell_1} \\
&\leq m^{-j} \sup_{k' \in \mathbb{Z}^d} \left\| \left(m^j (1 + \|M^j k - k'\|)^{-\delta} \right)_{k \in \mathbb{Z}^d} \right\|_{\ell_1} \left\| (d_k)_{k \in \mathbb{Z}^d} \right\|_{\ell_\infty} \\
&\lesssim m^{-j} \left\| (d_k)_{k \in \mathbb{Z}^d} \right\|_{\ell_\infty}.
\end{aligned}$$

Thus, (5.28) holds.

By applying the Riesz-Thorin Interpolation Theorem to (5.27) and (5.28), we obtain, for all $1 \leq \tau \leq \infty$,

$$\|T_j\|_{\ell_\tau \to \ell_\tau} \lesssim m^{-\frac{j}{\tau'}},$$

where $\frac{1}{\tau} + \frac{1}{\tau'} = 1$.

In order to address $j > 0$, we observe that, for $1 \leq \tau < \infty$, the operator $T_{-j} : \ell_{\tau'} \to \ell_{\tau'}$ is the dual matrix operator of $T_j : \ell_\tau \to \ell_\tau$. Thus,

$$\|T_j\|_{\ell_\tau \to \ell_\tau} = \|T_{-j}\|_{\ell_{\tau'} \to \ell_{\tau'}} \lesssim m^{\frac{j}{\tau}}.$$

Since $T_j : \ell_\infty \to \ell_\infty$ is the dual of $T_{-j} : \ell_1 \to \ell_1$, this inequality still holds for $\tau = \infty$, which concludes the proof. □

By following the lines of the proof in [BGN04], Lemma 5.3.2 implies the next Proposition.

Proposition 5.3.3. *Let M be isotropic, and let $1 \leq p < \infty$, $\delta > d$, and $s, s' \in \mathbb{N}$. Then the matrix operator*

$$(c_{j,k})_{j,k} \mapsto \left(\sum_{\substack{k \in \mathbb{Z}^d, \\ j \leq j'}} \frac{m^{(j-j')\left(\frac{s}{d} + \frac{1}{p}\right)} c_{j,k}}{(1 + \|k - M^{j-j'} k'\|)^\delta} + \sum_{\substack{k \in \mathbb{Z}^d, \\ j > j'}} \frac{m^{(j'-j)\left(\frac{s'}{d} + \frac{1}{p'}\right)} c_{j,k}}{(1 + \|k' - M^{j'-j} k\|)^\delta} \right)_{j',k'}$$

is bounded on $\ell_\tau(\mathbb{Z} \times \mathbb{Z}^d)$, for

$$p\left(\frac{s'}{d} + 1\right) > \tau > \begin{cases} \frac{d}{\delta}, & \text{for } \frac{s}{d} + \frac{1}{p} \geq \frac{\delta}{d}, \\ \left(\frac{s}{d} + \frac{1}{p}\right)^{-1}, & \text{for } 1 < \frac{s}{d} + \frac{1}{p} \leq \frac{\delta}{d}, \\ p(1 - \frac{s}{d}), & \text{for } \frac{s}{d} + \frac{1}{p} \leq 1. \end{cases}$$

The second fundamental ingredient for the proof of Theorem 5.3.1 is the following version of the Bramble-Hilbert Lemma, see [DL04]. Such inequalities are also known as Whitney estimates:

Theorem 5.3.4. *Given $\Omega \subset \mathbb{R}^d$ convex, $s \in \mathbb{N}$, and $1 \leq p \leq \infty$, let $f \in W^s(L_p(\Omega))$. Then there exists a polynomial $q \in \Pi_{s-1}$ such that*

$$|f-q|_{W^l(L_p(\Omega))} \lesssim \operatorname{diam}(\Omega)^{s-l}|f|_{W^s(L_p(\Omega))}, \quad l=0,\ldots,s,$$

where

$$|f|_{W^s(L_p)} := \sum_{|\beta|=s} \|\partial^\beta f\|_{L_p}$$

denotes the Sobolev semi-norm of order s.

The following proposition results by combining Proposition 5.3.3 with Theorem 5.3.4.

Proposition 5.3.5. *Let M be isotropic, $s, s' \in \mathbb{N}$, and suppose that compactly supported functions $f \in W^s(L_\infty(\mathbb{R}^d))$ and $g \in W^{s'}(L_\infty(\mathbb{R}^d))$ have s' and s vanishing moments, respectively. Given $1 \leq p < \infty$ and $1 = \frac{1}{p} + \frac{1}{p'}$, we consider the matrix operator*

$$T : (c_{j,k})_{j,k} \mapsto \left(\sum_{j\in\mathbb{Z},k\in\mathbb{Z}^d} \left\langle f^p_{j,k}, g^{p'}_{j',k'} \right\rangle c_{j,k} \right)_{j',k'}.$$

Then T is bounded on $\ell_\tau(\mathbb{Z}\times\mathbb{Z}^d)$ for any τ in the range

$$p\left(1+\frac{s'}{d}\right) > \tau > \begin{cases} \left(\frac{s}{d}+\frac{1}{p}\right)^{-1}, & \text{for } \frac{s}{d}+\frac{1}{p} \geq 1, \\ p\left(1-\frac{s}{d}\right), & \text{for } \frac{s}{d}+\frac{1}{p} \leq 1. \end{cases}$$

Proof. Fix $\delta > d$ sufficiently large. First, we address $j' \geq j$. Let $R > 0$ such that

$$\operatorname{supp}(g) \subset G := \{x \in \mathbb{R}^d : \|x\|_M \leq R\},$$

where $\|\cdot\|_M$ denotes the norm in (5.29). Then G is convex and $M^{j-j'}G \subset G$. According to the vanishing moments and the Hölder inequality, we obtain

$$\begin{aligned}
\left|\left\langle f^p_{j,k}, g^{p'}_{j',k'} \right\rangle\right| &= m^{\frac{j}{p}} m^{\frac{j'}{p'}} \int_{\mathbb{R}^d} f(M^{j-j'}x + M^{j-j'}k' - k)\overline{g(x)} m^{-j'} dx \\
&= m^{(j-j')\frac{1}{p}} \inf_{q\in\Pi_{s-1}} \int_G \left(f(M^{j-j'}x + M^{j-j'}k' - k) - q(x) \right) \overline{g(x)} dx \\
&\leq m^{(j-j')\frac{1}{p}} \inf_{q\in\Pi_{s-1}} \left\| f(M^{j-j'}\cdot + M^{j-j'}k' - k) - q(\cdot) \right\|_{L_\infty(G)} \|g\|_{L_1(G)}.
\end{aligned}$$

The space Π_{s-1} is affine invariant, i.e., $q \in \Pi_{s-1}$ yields $q(A\cdot + t) \in \Pi_{s-1}$, for all $A \in \mathbb{R}^{d\times d}$ and $t \in \mathbb{R}^d$. Thus, Theorem 5.3.4 with $l = 0$ implies

$$\begin{aligned}
\left|\left\langle f^p_{j,k}, g^{p'}_{j',k'} \right\rangle\right| &\lesssim m^{(j-j')\frac{1}{p}} \inf_{q\in\Pi_{s-1}} \|f-q\|_{L_\infty(M^{j-j'}G + M^{j-j'}k'-k)} \\
&\lesssim m^{(j-j')\frac{1}{p}} \operatorname{diam}(M^{j-j'}G)^s |f|_{W^s(L_\infty(M^{j-j'}G+M^{j-j'}k'-k))} \\
&\lesssim m^{(j-j')\frac{1}{p}} m^{(j-j')\frac{s}{d}} |f|_{W^s(L_\infty(G+M^{j-j'}k'-k))}.
\end{aligned}$$

Since f is compactly supported, there exists $r > 0$ such that, for all $v \in \mathbb{R}^d$ with $\|v\| \geq r$, the intersection $(G + v) \cap \operatorname{supp}(f)$ is empty. Hence, the Sobolev semi-norm can be estimated by

$$|f|_{W^s(L_\infty(G+M^{j-j'}k'-k))} \leq \begin{cases} |f|_{W^s(L_\infty(\mathbb{R}^d))}, & \text{for } \|M^{j-j'}k' - k\| < r, \\ 0, & \text{for } \|M^{j-j'}k' - k\| \geq r. \end{cases}$$

This provides the final inequalities

$$\begin{aligned} \left|\left\langle f^p_{j,k}, g^{p'}_{j',k'}\right\rangle\right| &\lesssim m^{(j-j')(\frac{s}{d}+\frac{1}{p})} |f|_{W^s(L_\infty(\mathbb{R}^d))} \left(\frac{1+r}{1+\|M^{j-j'}k'-k\|}\right)^\delta \\ &\lesssim \frac{m^{(j-j')(\frac{s}{d}+\frac{1}{p})}}{(1+\|M^{j-j'}k'-k\|)^\delta}. \end{aligned}$$

Next, we address $j > j'$. Following the lines above with interchanged roles of f and g, we obtain

$$\left|\left\langle f^p_{j,k}, g^{p'}_{j',k'}\right\rangle\right| \lesssim \frac{m^{(j'-j)(\frac{s'}{d}+\frac{1}{p'})}}{(1+\|M^{j'-j}k-k'\|)^\delta}.$$

By applying Proposition 5.3.3, the operator T is bounded on ℓ_τ. □

Proposition 5.3.5 addresses single f and g. In order to consider a finite number of functions as in Theorem 5.3.1, one applies norm coherences between $\ell_\tau(E \times \mathbb{Z} \times \mathbb{Z}^d)$ and $\ell_\tau(\mathbb{Z} \times \mathbb{Z}^d)$. We omit the detailed elaboration.

5.3.2 Hilbertian Dictionaries

Given a pair of biorthogonal wavelet bases, then, under the assumptions of Theorem 5.2.6, for $f \in \dot{B}^\alpha$, the series expansion

$$\sum_{\lambda\in\Lambda} \left\langle f, \widetilde{\psi}^{p'}_\lambda \right\rangle \psi^p_\lambda$$

converges towards f in $L_p(\mathbb{R}^d)$. This subsection provides some fundamentals, in order to generalize this statement regarding wavelet bi-frames. Given a wavelet system $\{\psi_\lambda : \lambda \in \Lambda\}$, we derive a classical decay condition on the sequence $(c_\lambda)_{\lambda\in\Lambda}$ such that

$$\sum_{\lambda\in\Lambda} c_\lambda \psi^p_\lambda \tag{5.30}$$

converges in $L_p(\mathbb{R}^d)$.

Naturally, convergence problems as in (5.30) also arise in more abstract settings. In order to point out the key ingredients of its solution, we study the problem in a general framework. Let X be a Banach space. A countable subset $\mathcal{D}$ in X is called a *dictionary* if its elements are normalized in the sense of $\|g\|_X \sim 1$, for all $g \in \mathcal{D}$. Although we mainly think of complete dictionaries, i.e., their linear span is dense in X, we do not suppose its completeness in advance.

In order to obtain a sufficient variety of decay conditions, we recall the following family of sequence spaces:

Definition 5.3.6. For $0 < p < \infty$, $0 < q \leq \infty$ and a countable index set $\mathcal{K}$, the *Lorentz space* $\ell_{p,q}(\mathcal{K})$ is the collection of bounded sequences $(c_\kappa)_{\kappa\in\mathcal{K}}$ satisfying $\|(c_\kappa)_{\kappa\in\mathcal{K}}\|_{\ell_{p,q}} < \infty$, where

$$\|(c_\kappa)_{\kappa\in\mathcal{K}}\|_{\ell_{p,q}} := \begin{cases} \left(\sum_{j=1}^{\infty}(j^{\frac{1}{p}}c_j^*)^q\frac{1}{j}\right)^{\frac{1}{q}}, & \text{for } 0 < q < \infty, \\ \sup_{j\geq 1}(j^{\frac{1}{p}}c_j^*), & \text{for } q = \infty, \end{cases} \tag{5.31}$$

while $(c_j^*)_{j\in\mathbb{N}}$ denotes a decreasing rearrangement of $(|c_\kappa|)_{\kappa\in\mathcal{K}}$.

Note that $\ell_p(\mathcal{K}) = \ell_{p,p}(\mathcal{K})$. Hence, Lorentz spaces refine the scale of ℓ_p spaces. Since $(c_j^*)_{j\in\mathbb{N}}$ is a decreasing sequence, Cauchy's condensation test provides the norm equivalences

$$\|(c_\kappa)_{\kappa\in\mathcal{K}}\|_{\ell_{p,q}} \sim \begin{cases} \left(\sum_{j=0}^{\infty}(2^{\frac{j}{p}}c_{2^j}^*)^q\right)^{\frac{1}{p}}, & \text{for } 0 < q < \infty, \\ \sup_{j\geq 0}\{2^{\frac{j}{p}}c_{2^j}^*\}, & \text{for } q = \infty. \end{cases} \tag{5.32}$$

According to the results of Section 4.3 in [Lin05], the number 2 in (5.32) can be replaced by any fixed integer greater than 1. It is often more convenient to work with (5.32) than with the original norm (5.31). For instance, by applying (5.32), the ideas at the end of Chapter 2 in [DL93] provide the following continuous embeddings, for $0 < \varepsilon$,

$$\ell_{p+\varepsilon,\infty}(\mathcal{K}) \hookrightarrow \ell_{p,q}(\mathcal{K}), \tag{5.33}$$

$$\ell_{p,q}(\mathcal{K}) \hookrightarrow \ell_{p,q+\varepsilon}(\mathcal{K}) \hookrightarrow \ell_{p,\infty}(\mathcal{K}). \tag{5.34}$$

Hence, similar to Besov spaces, the index q is only of minor interest. Recall that, for $f \in L_p(\mathbb{R}^d)$, the modulus of smoothness $\omega_l(f,t)_{L_p}$ is decreasing as t decreases. Then (5.10) and (5.32) provide

$$|f|_{B_q^s(L_p)} \sim \left\|\left(\omega_l(f,\tfrac{1}{j})_{L_p}\right)_{j\in\mathbb{N}}\right\|_{\ell_{\frac{1}{s},q}}.$$

Following [BGN04, GN04], we introduce a specific family of dictionaries:

Definition 5.3.7. A dictionary $\mathcal{D} = \{g_\kappa : \kappa \in \mathcal{K}\}$ in a Banach space X is called $\ell_{p,q}(\mathcal{K})$-*hilbertian* if the synthesis-type operator

$$F : \ell_{p,q}(\mathcal{K}) \to X, \quad (c_\kappa)_{\kappa\in\mathcal{K}} \mapsto \sum_{\kappa\in\mathcal{K}} c_\kappa g_\kappa$$

is well-defined and bounded.

For $q = 1$, hilbertian dictionaries are characterized in the following Proposition.

Proposition 5.3.8. *Let $\mathcal{D} = \{g_\kappa : \kappa \in \mathcal{K}\}$ be a dictionary in a Banach space X and $1 \leq p < \infty$. Then the following properties are equivalent:*

(i) *$\mathcal{D}$ is $\ell_{p,1}(\mathcal{K})$-hilbertian.*

(ii) *For all index sets $\mathcal{K}_N \subset \mathcal{K}$ of cardinality N and every choice of signs*

$$\left\|\sum_{\kappa\in\mathcal{K}_N} \pm g_\kappa\right\|_X \lesssim N^{\frac{1}{p}}.$$

(iii) *For all index sets $\mathcal{K}_N \subset \mathcal{K}$ of cardinality N and every sequence $(d_\kappa)_{\kappa\in\mathcal{K}_N} \in \ell(\mathcal{K}_N)$*

$$\left\| \sum_{\kappa\in\mathcal{K}_N} d_\kappa g_\kappa \right\|_X \lesssim N^{\frac{1}{p}} \max_{\kappa\in\mathcal{K}_N} |d_\kappa|. \tag{5.35}$$

The equivalence between (i) and (ii) has already been derived in [GN04]. We extend the result to condition (iii).

Proof. Obviously, (iii) implies (ii). Let us show that (i) implies (iii): for a given sequence $(d_\kappa)_{\kappa\in\mathcal{K}_N} \in \ell(\mathcal{K}_N)$, we define a second sequence $(c_\kappa)_{\kappa\in\mathcal{K}}$ by

$$c_\kappa := \begin{cases} d_\kappa, & \text{for } \kappa \in \mathcal{K}_N, \\ 0, & \text{otherwise.} \end{cases}$$

Since $(c_\kappa)_{\kappa\in\mathcal{K}} \in \ell_{p,1}(\mathcal{K})$, applying (i) yields

$$\begin{aligned} \left\| \sum_{\kappa\in\mathcal{K}_N} d_\kappa g_\kappa \right\|_X &\lesssim \|(c_\kappa)_{\kappa\in\mathcal{K}}\|_{\ell_{p,1}} = \sum_{j=1}^{\infty} j^{\frac{1}{p}-1} c_j^* \\ &\le \max_{\kappa\in\mathcal{K}_N} |c_\kappa| \sum_{j=1}^{N} j^{\frac{1}{p}-1} = \max_{\kappa\in\mathcal{K}_N} |d_\kappa| \, N^{\frac{1}{p}} \frac{1}{N} \sum_{j=1}^{N} \left(\frac{j}{N}\right)^{\frac{1}{p}-1}. \end{aligned}$$

A Riemann sum argument provides

$$\frac{1}{N}\sum_{j=1}^{N} \left(\frac{j}{N}\right)^{\frac{1}{p}-1} \le \int_0^1 x^{\frac{1}{p}-1} dx = p.$$

This concludes the proof. □

In the sequel, we establish that a compactly supported wavelet system, properly normalized in $L_p(\mathbb{R}^d)$, is $\ell_{p,1}$-hilbertian. In order to derive the result, the following statement about overlapping supports is helpful:

Lemma 5.3.9. *Let a finite number of compactly supported functions $\psi^{(\mu)}$, $\mu = 1, \dots, n$, be given. Then their dilates and shifts satisfy the following overlapping condition:*

(a) $\left|\operatorname{supp}\left(\psi_{j,k}^{(\mu)}\right)\right| \lesssim m^{-j}$.

(b) *Let μ and k be fixed. Then, for all $j \in \mathbb{Z}^d$,*

$$\operatorname{card}\left\{(\nu, l) : \operatorname{supp}\left(\psi_{j,k}^{(\mu)}\right) \cap \operatorname{supp}\left(\psi_{j,l}^{(\nu)}\right) \neq \emptyset\right\} \lesssim 1.$$

The result of Lemma 5.3.9 is well-known for dyadic dilation, cf. [Coh03]. Essentially, its proof does not depend on the dilation, but see Appendix A.2 for the extension to general scalings.

The following lemma is a standard component in nonlinear approximation theory for dyadic dilation, cf. [Coh03, DeV98]. A proof for general isotropic scalings can be found in [Lin05]. Note that it is stated under an additional basis assumption. An analysis of the proof yields that the assumption is not necessary. One only uses the overlapping conditions of Lemma 5.3.9.

Lemma 5.3.10. *Let $\psi^{(\mu)}$, $\mu = 1,\dots,n$, be compactly supported functions in $L_\infty(\mathbb{R}^d)$. Assume $1 \le p < \infty$. Then, for all $\Lambda_N \subset \{1,\dots,n\} \times \mathbb{Z} \times \mathbb{Z}^d$ of cardinality N and every sequence $(d_\lambda)_{\lambda\in\Lambda_N} \in \ell(\Lambda_N)$*

$$\left\| \sum_{\lambda\in\Lambda_N} d_\lambda \psi_\lambda \right\|_{L_p} \lesssim N^{\frac{1}{p}} \max_{\lambda\in\Lambda_N} \|d_\lambda\psi_\lambda\|_{L_p}. \tag{5.36}$$

Actually, (5.36) is just a rephrasing of (5.35) involving the $L_p(\mathbb{R}^d)$-normalization:

Corollary 5.3.11. *Let $\psi^{(\mu)}$, $\mu = 1,\dots,n$, be compactly supported functions in $L_\infty(\mathbb{R}^d)$ and $1 \le p < \infty$. Then, with $\Lambda = \{1,\dots,n\} \times \mathbb{Z} \times \mathbb{Z}^d$, the L_p-normalized wavelet system*

$$\{\psi_\lambda^p : \lambda \in \Lambda\}$$

is an $\ell_{p,1}(\Lambda)$-hilbertian dictionary in $L_p(\mathbb{R}^d)$.

Proof. Given $\Lambda_N \subset \Lambda$, card $(\Lambda_N) = N$, and $(d_\lambda)_{\lambda\in\Lambda_N} \in \ell(\Lambda_N)$, define $(c_\lambda)_{\lambda\in\Lambda_N} \in \ell(\Lambda_N)$ such that $c_\lambda\psi_\lambda = d_\lambda\psi_\lambda^p$, for $\lambda \in \Lambda_N$. Then Lemma 5.3.10 yields

$$\begin{aligned}\left\| \sum_{\lambda\in\Lambda_N} d_\lambda \psi_\lambda^p \right\|_{L_p} &= \left\| \sum_{\lambda\in\Lambda_N} c_\lambda \psi_\lambda \right\|_{L_p} \lesssim N^{\frac{1}{p}} \max_{\lambda\in\Lambda_N} \|c_\lambda\psi_\lambda\|_{L_p} \\ &= N^{\frac{1}{p}} \max_{\lambda\in\Lambda_N} \|d_\lambda\psi_\lambda^p\|_{L_p} \lesssim N^{\frac{1}{p}} \max_{\lambda\in\Lambda_N} |d_\lambda|.\end{aligned}$$

The last inequality holds because ψ_λ^p is normalized in $L_p(\mathbb{R}^d)$. By applying Proposition 5.3.8, we conclude the proof. □

According to Corollary 5.3.11, the series $\sum_{\lambda\in\Lambda} c_\lambda\psi_\lambda^p$ converges in $L_p(\mathbb{R}^d)$ if $(c_\lambda)_{\lambda\in\Lambda}$ is contained in $\ell_{p,1}(\Lambda)$. In order to consider wavelet bi-frame expansions

$$f = \sum_{\lambda\in\Lambda} \left\langle f, \widetilde{\psi}_\lambda^{p'} \right\rangle \psi_\lambda^p \tag{5.37}$$

in $L_p(\mathbb{R}^d)$, there still remain two problems. First, we have to verify that the coefficient sequence $\left(\left\langle f, \widetilde{\psi}_\lambda^{p'} \right\rangle\right)_{\lambda\in\Lambda}$ is contained in $\ell_{p,1}(\Lambda)$. Then the right-hand side of (5.37) converges in $L_p(\mathbb{R}^d)$. Second, we have to verify that the series converges towards f. Both problems are addressed in the following Subsection 5.3.3.

5.3.3 Norm Equivalences for Homogeneous Besov Spaces

In this subsection, we finally derive the characterization of the homogeneous Besov space by wavelet bi-frames with general isotropic scalings. The following theorem extends dyadic results in [BGN04].

Theorem 5.3.12. *Given $1 < p < \infty$, $\frac{1}{p} + \frac{1}{p'} = 1$, let*

$$X(\{\psi^{(1)},\dots,\psi^{(n)}\}), \quad X(\{\widetilde{\psi}^{(1)},\dots,\widetilde{\psi}^{(n)}\})$$

be a compactly supported wavelet bi-frame. In addition, suppose that

$$X(\{\eta^{(1)},\dots,\eta^{(m-1)}\}),\quad X(\{\widetilde{\eta}^{(1)},\dots,\widetilde{\eta}^{(m-1)}\})$$

is a pair of compactly supported biorthogonal wavelet bases. Given $s, s' \in \mathbb{N}$, then let, for $\mu = 1,\dots,n$ and $\nu = 1,\dots,m-1$,

$$\psi^{(\mu)},\eta^{(\nu)} \in W^s(L_\infty(\mathbb{R}^d)) \quad \text{and} \quad \widetilde{\psi}^{(\mu)},\widetilde{\eta}^{(\nu)} \in W^{s'}(L_\infty(\mathbb{R}^d))$$

have s' and s vanishing moments, respectively. If the pair of biorthogonal wavelet bases characterizes $\dot{B}^\alpha$ in the sense of Theorem 5.2.6, then we have, for α in the range

$$0 < \alpha < \begin{cases} s, & \text{for } \frac{s}{d} + \frac{1}{p} \geq 1, \\ \frac{s}{p\left(1-\frac{s}{d}\right)}, & \text{for } \frac{s}{d} + \frac{1}{p} \leq 1, \end{cases}$$

and for all $f \in \dot{B}^\alpha$, that

$$f = \sum_{\lambda \in \Lambda} \left\langle f, \widetilde{\psi}_\lambda^{p'} \right\rangle \psi_\lambda^p \tag{5.38}$$

holds in $\dot{B}^\alpha$ (and so in $L_p(\mathbb{R}^d)$) and

$$\|f\|_{\dot{B}^\alpha} \sim \left\| \left(\left\langle f, \widetilde{\psi}_\lambda^{p'} \right\rangle \right)_{\lambda \in \Lambda} \right\|_{\ell_\tau(\Lambda)}, \tag{5.39}$$

where $\frac{1}{\tau} = \frac{\alpha}{d} + \frac{1}{p}$ and $\Lambda = \{1,\dots,n\} \times \mathbb{Z} \times \mathbb{Z}^d$.

Remark 5.3.13. Theorem 5.3.12 requires the existence of a biorthogonal reference wavelet system, which already characterizes the Besov space. In the dyadic setting of [BGN04], this assumptions is not explicitly mentioned since one can simply choose tensor products of sufficiently smooth orthonormal Daubechies wavelets or the Meyer wavelet, see [Dau92]. As far as we know, it is still an unanswered question, whether, for each isotropic dilation matrix, one can find families of arbitrarily smooth compactly supported pairs of biorthogonal wavelet bases. Hence, we had to formulate the existence of a reference system as an assumption in Theorem 5.3.12. We should point out, that, for many nondyadic scalings, these families exist, cf. [Der99, JRS99], and the characterization is applicable. Finally, we conjecture that the above question has a positive answer. Since one allows for arbitrarily large support sizes, we expect that the overwhelming majority of isotropic dilation matrices has such biorthogonal reference wavelets.

One may worry that large supports of pairs of biorthogonal wavelet bases also force large constants in the biorthogonal characterization. Then the norm equivalence (5.39) may also inherit very large constants since they depend on those for the biorthogonal wavelets as well as on the localization technique itself. Hence, Theorem 5.3.12 is only a qualitative result. Nevertheless, we expect that our method is far-off from providing optimal constants. The true ones should be much better, and this point of view is supported by the successful application of the norm equivalences (5.39) to image denoising in Chapter 7.

For preparation, we need the following simple lemma, see Appendix A.2 for the proof.

Lemma 5.3.14. *Let $1 \leq q < p \leq \infty$ and let $f_n \in L_p(\mathbb{R}^d) \cap L_q(\mathbb{R}^d)$, $n \in \mathbb{N}$, converge to f in $L_p(\mathbb{R}^d)$ and to g in $L_q(\mathbb{R}^d)$. Then $f = g$ up to a set of measure zero.*

Proof of Theorem 5.3.12. Let $f \in \dot{B}^\alpha$ and $\Lambda' = \{1, \ldots, m-1\} \times \mathbb{Z} \times \mathbb{Z}^d$, then

$$f = \sum_{\lambda' \in \Lambda'} \left\langle f, \widetilde{\eta}^{p'}_{\lambda'} \right\rangle \eta^p_{\lambda'}$$

holds in $L_p(\mathbb{R}^d)$ and

$$\|f\|_{\dot{B}^\alpha} \sim \left\| \left(\left\langle f, \widetilde{\eta}^{p'}_{\lambda'} \right\rangle \right)_{\lambda' \in \Lambda'} \right\|_{\ell_\tau}. \tag{5.40}$$

For $\frac{s}{d} + \frac{1}{p} \geq 1$, we have

$$\frac{1}{\tau} = \frac{\alpha}{d} + \frac{1}{p} < \frac{s}{d} + \frac{1}{p}.$$

Hence,

$$p > \tau > \left(\frac{s}{d} + \frac{1}{p} \right)^{-1},$$

and τ is in the admissible range of Theorem 5.3.1. For $\frac{s}{d} + \frac{1}{p} \leq 1$, we have

$$\begin{aligned} \frac{1}{\tau} = \frac{\alpha}{d} + \frac{1}{p} &< \frac{s}{p(d-s)} + \frac{1}{p} \\ = \frac{d}{p(d-s)} &= \frac{1}{p\left(1 - \frac{s}{d}\right)}. \end{aligned}$$

Thus, Theorem 5.3.1 can be applied in both cases. Then we obtain

$$\begin{aligned} \left\| \left(\left\langle f, \widetilde{\psi}^{p'}_{\lambda} \right\rangle \right)_{\lambda \in \Lambda} \right\|_{\ell_\tau} &= \left\| \left(\sum_{\lambda' \in \Lambda'} \left\langle f, \widetilde{\eta}^{p'}_{\lambda'} \right\rangle \left\langle \eta^p_{\lambda'}, \widetilde{\psi}^{p'}_{\lambda} \right\rangle \right)_{\lambda \in \Lambda} \right\|_{\ell_\tau} && (5.41) \\ &\lesssim \left\| \left(\left\langle f, \widetilde{\eta}^{p'}_{\lambda'} \right\rangle \right)_{\lambda' \in \Lambda'} \right\|_{\ell_\tau}. && (5.42) \end{aligned}$$

With (5.40), this implies

$$\left\| \left(\left\langle f, \widetilde{\psi}^{p'}_{\lambda} \right\rangle \right)_{\lambda \in \Lambda} \right\|_{\ell_\tau} \lesssim \|f\|_{\dot{B}^\alpha}. \tag{5.43}$$

For the reverse estimate, wavelet bi-frame and biorthogonal wavelets change roles in the localization process. First, we establish (5.38). According to Corollary 5.3.11, the primal bi-frame wavelets

$$\left\{ \psi^p_\lambda : \lambda \in \Lambda \right\}$$

are $\ell_{p,1}$-hilbertian. Hence, the synthesis-type operator

$$F : \ell_{p,1} \to L_p(\mathbb{R}^d), \quad (d_\lambda)_{\lambda \in \Lambda} \mapsto \sum_{\lambda \in \Lambda} d_\lambda \psi^p_\lambda$$

is well-defined and bounded. By applying (5.43), the analysis-type operator

$$\widetilde{F}^* : \dot{B}^\alpha \to \ell_\tau, \quad f \mapsto \left(\left\langle f, \widetilde{\psi}^{p'}_{\lambda'} \right\rangle \right)_{\lambda'}$$

$$\begin{array}{ccc} \dot{B}^\alpha & \xrightarrow{\widetilde{F}^*} & \ell_\tau \\ \Big\downarrow i & & \Big\downarrow i \\ L_p(\mathbb{R}^d) & \xleftarrow{F} & \ell_{p,1} \end{array}$$

Figure 5.4: Mapping diagram for analysis- and synthesis-type operators, where i denotes the canonical embedding

is bounded (the notation may only remind of the original analysis operator (2.2) on Hilbert spaces. The present operator $\widetilde{F}^*$ is neither considered as any adjoint on Hilbert spaces nor any dual operator on Banach spaces). Note that the embedding

$$\ell_\tau(\Lambda) \hookrightarrow \ell_{p,1}(\Lambda)$$

holds. Then in order to establish that the diagram in Figure 5.4 commutes, we consider the bounded operator

$$F\widetilde{F}^* : \dot{B}^\alpha \to L_p(\mathbb{R}^d)$$

more closely.

Since $\dot{B}^\alpha$ is contained in $L_p(\mathbb{R}^d)$, cf. (5.20), Lemma 5.3.14 and the bi-frame expansion (2.9) in $L_2(\mathbb{R}^d)$ imply that $F\widetilde{F}^*$ is the identity on $\dot{B}^\alpha \cap L_2(\mathbb{R}^d)$. According to the results of Chapter 1 in [RS96], the intersection $\dot{B}^\alpha \cap L_2(\mathbb{R}^d)$ is dense in $\dot{B}^\alpha$. Hence, the continuity of $F\widetilde{F}^*$ finally yields that (5.38) holds in $L_p(\mathbb{R}^d)$, and the diagram in Figure 5.4 commutes.

By following (5.41), (5.42) with interchanged roles of $\widetilde{\psi}$, $\widetilde{\eta}$ as well as η replaced by ψ, we obtain the reverse estimate of (5.43).

We still have to address the convergence of (5.38) in $\dot{B}^\alpha$. Since primal and dual wavelets are not required to be biorthogonal, the partial sums of the series (5.38) may have different wavelet coefficients than the $\langle f, \widetilde{\psi}^{p'}_\lambda \rangle$, $\lambda \in \Lambda_N$, which appear in the sum itself, i.e., for a subset $\Lambda_N \subset \Lambda$ of cardinality N, we could face

$$\widetilde{F}^*\Big(\sum_{\lambda \in \Lambda_N} \big\langle f, \widetilde{\psi}^{p'}_\lambda \big\rangle \psi^p_\lambda \Big) \neq \Big(\big\langle f, \widetilde{\psi}^{p'}_\lambda \big\rangle \Big)_{\lambda \in \Lambda_N}.$$

Thus, the norm equivalence (5.39) does not directly imply that the partial sums constitute a Cauchy sequence. Nevertheless, the matrix operator $\Big(\langle \psi^p_\lambda, \widetilde{\psi}^{p'}_{\lambda'} \rangle \Big)_{\lambda,\lambda'}$ is bounded on $\ell_\tau(\Lambda)$, see Theorem 5.3.1. By applying this additional ingredient, the norm equivalences provide the Cauchy property. According to the completeness of $\dot{B}^\alpha$, we can conclude the proof. □

In Chapters 3 and 4, we obtained several smooth wavelet bi-frames with isotropic dilation and a high number of vanishing moments. Then, according to Theorem 5.3.12, they characterize Besov spaces. Since they have sufficient vanishing moments, as for biorthogonal wavelets, the range of Besov spaces is only restricted by their smoothness:

Example 5.3.15. Let $1 < p < \infty$. Then the wavelet bi-frame Laplace (3-2) as well as its reduced counterpart Laplace (3-2)$_\mathrm{R}$ is contained in $W^3(L_\infty(\mathbb{R}^2))$ and all wavelets have at least four vanishing moments. Note that there exists a pair of biorthogonal wavelet bases, which can play the role of the reference system in Theorem 5.3.12, cf. [Der99, JRS99] (recall that the support sizes are allowed to be as large as necessary). Thus, the bi-frame characterizes the bivariate Besov spaces $\dot{B}^{\alpha}$, for all $0 < \alpha < 3$.

We explicitly establish the following particular result since it is required in Chapter 7.

Example 5.3.16. Laplace (2-2) is contained in $W^2(L_\infty(\mathbb{R}^2))$ and each wavelet has at least four vanishing moments. Thus, it characterizes $\dot{B}^1$ with $d = p = 2$.

We even obtain the characterization for the complete range of multivariate Besov spaces $\dot{B}^{\alpha}$, $\alpha > 0$, since the dilation matrix of the Checkerboard wavelet bi-frames in Section 3.3.1 and Example 4.3.1 allows for arbitrarily smooth pairs of compactly supported biorthogonal wavelet bases.

Example 5.3.17. Let $1 < p < \infty$. For each fixed $0 < s \in \mathbb{N}$, there exists a sufficiently large N, such that the multivariate wavelet bi-frames Checkerboard (N) and $(N)_{\mathrm{R}}$ are contained in $W^s(L_\infty(\mathbb{R}^d))$, and all wavelets have s vanishing moments. Thus, they characterize the multivariate Besov spaces $\dot{B}^{\alpha}$, for all $0 < \alpha < s$.

Finally, we apply Theorem 5.3.12 to our bi-frames Box Spline $(N)_{\mathrm{R}}$ in Example 4.3.3. Since it involves dyadic dilation, the following example can also be derived by the results in [BGN04].

Example 5.3.18. Let $1 < p < \infty$. Then the wavelet bi-frame Box Spline $(4)_{\mathrm{R}}$ is contained in $W^5(L_\infty(\mathbb{R}^2))$ and all wavelets have at least 8 vanishing moments. Hence, it characterizes the bivariate Besov spaces $\dot{B}^{\alpha}$, for all $0 < \alpha < 5$.

Chapter 6

N-Term Approximation by Wavelet Bi-Frames

In order to analyze a function, we decompose it into simple building blocks. These blocks also provide a series expansion, which reconstructs the original function. In computational algorithms, the series has to be replaced by a finite sum. Hence, we must approximate from N terms. There arise two fundamental problems, which have to be solved. First, let the approximation class essentially collect all functions, whose best choice of N terms yields a specific rate of approximation. It is important to express the approximation class in terms of classical function spaces since the class serves as a benchmark in order to evaluate different selections of N terms. Second, in practical algorithms, we require a realization of the best N-term approximation, i.e., we must look for a simple rule of the selection of N particular terms such that they provide the same rate of approximation as the best N-term approximation.

In wavelet theory, one approximates functions from dilates and shifts. For dyadic orthonormal wavelets, at least up to a certain rate, the approximation class equals a Besov space and thresholding the coefficients of the series expansion realizes the best N-term approximation, cf. [Coh03, DeV98]. The results require certain smoothness and vanishing moments of the wavelets as well as a linear independence condition on the underlying refinable function. In [Lin05], Lindemann generalized the dyadic results regarding biorthogonal wavelet bases with isotropic dilation. Borup, Gribonval, and Nielsen address wavelet bi-frames in [BGN04]. However, their results are restricted to dyadic dilation. In the present chapter, we try to extend their results to more general scalings.

First, we recall the basic elements of best N-term approximation in Banach spaces, cf. [DL93]. In order to characterize the approximation class, one has to establish so-called matching Jackson and Bernstein inequalities. They imply that the approximation space equals a so-called interpolation space. Fortunately, interpolation is well-studied, and, in many particular situations, these classes can be identified with classical smoothness spaces, which yields the final characterization of the approximation class.

For wavelet bi-frames with idempotent scaling, we establish matching Jackson and Bernstein inequalities. Hence, their approximation classes are interpolation spaces. Since the arising interpolation spaces coincide at least for certain parameters with Besov spaces, we solved the first problem mentioned above. Facing the second problem, we derive that the best N-term approximation rate can be realized by thresholding the wavelet bi-frame expansion.

The limitation to idempotent scalings is not too restrive since our wavelet bi-frames in Chapters 3 and 4 are included. In the remainder of the chapter, we verify that they satisfy the assumptions of the Jackson and Bernstein inequalities.

6.1 Best N-Term Appoximation

6.1.1 The Approximation Class

Given some dictionary $\mathcal{D}$ in a Banach space X, let $\Sigma_N(\mathcal{D})$ be the collection of all linear combinations of at most N elements of $\mathcal{D}$. For any given $f \in X$,

$$\sigma_N(f,\mathcal{D})_X := \operatorname{dist}(f,\Sigma_N(\mathcal{D}))_X$$

is called the *error of best N-term approximation.* In order to approximate elements in X from $\Sigma_N(\mathcal{D})$, it is important to determine those $f \in X$ providing the *approximation rate* α, i.e.,

$$\sigma_N(f,\mathcal{D})_X \lesssim N^{-\alpha}, \quad \text{for all } N \in \mathbb{N},$$

where the constant may depend on f. This question leads to the following definition:

Definition 6.1.1. Let $\mathcal{D}$ be a dictionary in some Banach space X. For $0 < s < \infty$, $0 < q \leq \infty$, the *approximation class* $\mathcal{A}^s_q(X,\mathcal{D})$ is the collection of all $f \in X$ such that

$$|f|_{\mathcal{A}^s_q(X,\mathcal{D})} := \begin{cases} \left(\sum_{N=1}^{\infty} (N^s\sigma_N(f,\mathcal{D})_X)^q \frac{1}{N}\right)^{\frac{1}{q}}, & \text{for } 0<q<\infty, \\ \sup_{N\geq 1}(N^s\sigma_N(f,\mathcal{D})_X), & \text{for } q=\infty, \end{cases} \tag{6.1}$$

is finite. It is quasi-normed by

$$\|f\|_{\mathcal{A}^s_q(X,\mathcal{D})} := \|f\|_X + |f|_{\mathcal{A}^s_q(X,\mathcal{D})}. \tag{6.2}$$

If we choose $q=\infty$, then the space $\mathcal{A}^s_\infty(X,\mathcal{D})$ precisely consists of all f in X having approximation rate s. For $0<q<\infty$, membership in $\mathcal{A}^s_q(X,\mathcal{D})$ means a slightly stronger condition: since $\sigma_N(f,\mathcal{D})_X$ is decreasing in N, the Lorentz space embeddings (5.33) and (5.34) provide, for all $\varepsilon > 0$,

$$\mathcal{A}^{s+\varepsilon}_\infty(X,\mathcal{D}) \hookrightarrow \mathcal{A}^s_q(X,\mathcal{D}), \tag{6.3}$$

$$\mathcal{A}^s_q(X,\mathcal{D}) \hookrightarrow \mathcal{A}^s_{q+\varepsilon}(X,\mathcal{D}) \hookrightarrow \mathcal{A}^s_\infty(X,\mathcal{D}), \tag{6.4}$$

see also Chapter 7 in [DL93]. According to (6.4), we obtain, for all $f \in \mathcal{A}^s_q(X,\mathcal{D})$,

$$\sigma_N(f,\mathcal{D})_X \lesssim N^{-s}\|f\|_{\mathcal{A}^s_\infty} \lesssim N^{-s}\|f\|_{\mathcal{A}^s_q}. \tag{6.5}$$

Hence, in (6.5), we explicitly establish that membership in $\mathcal{A}^s_q(X,\mathcal{D})$ is slightly stronger than having approximation rate s.

Remark 6.1.2. Originally, we are interested in $\mathcal{A}^s_\infty(X,\mathcal{D})$. In approximation theory, one is often not able to characterize this spaces, but one can describe the slightly smaller space $\mathcal{A}^s_q(X,\mathcal{D})$ at least for a certain q. Since the differences are only marginal and both classes correspond to approximation rate s, this characterization is sufficient for most purposes.

Next, we recall some results of best N-term approximation in $L_p(\mathbb{R}^d)$, where the dictionary consists of wavelets. The following theorem addresses biorthogonal wavelet bases with isotropic dilation, cf. [Lin05]:

Theorem 6.1.3. *Let M be an isotropic dilation matrix, and let $X(\{\psi^{(1)},\dots,\psi^{(m-1)}\})$, $X(\{\widetilde{\psi}^{(1)},\dots,\widetilde{\psi}^{(n)}\})$ be a pair of compactly supported biorthogonal wavelet bases with underlying primal refinable function $\varphi \in W^s(L_\infty(\mathbb{R}^d))$. Given $1 < p < \infty$, then for $0 < \alpha < s$,*

$$\mathcal{A}_\tau^{\frac{\alpha}{d}}\left(L_p, X(\{\psi^{(1)},\dots,\psi^{(m-1)}\})\right) = \dot{B}^\alpha,$$

where $\frac{1}{\tau} = \frac{\alpha}{d} + \frac{1}{p}$.

For compactly supported wavelet bi-frames, the results in [BGN04] imply the following theorem. Note the restriction to dyadic dilation, and see Section 5.2 for linear independence on open sets:

Theorem 6.1.4. *Given the matrix $M = 2\mathcal{I}_d$ and $1 < p < \infty$, let $X(\{\psi^{(1)},\dots,\psi^{(n)}\})$, $X(\{\widetilde{\psi}^{(1)},\ \dots,\widetilde{\psi}^{(n)}\})$ be a compactly supported wavelet bi-frame with compactly supported refinable functions $\varphi \in \mathcal{C}^s(\mathbb{R}^d)$, $s \in \mathbb{N}$, and $\widetilde{\varphi} \in \mathcal{C}^1(\mathbb{R}^d)$, respectively. Moreover, let φ have linearly independent integer shifts on $(0,1)^d$. If all $\widetilde{\psi}^{(\mu)}$, $\mu = 1,\dots,n$, have s vanishing moments, then for*

$$0 < \alpha < \begin{cases} s, & \text{for } \frac{s}{d} + \frac{1}{p} \geq 1, \\ \frac{s}{p(1-\frac{s}{d})}, & \text{for } \frac{s}{d} + \frac{1}{p} \leq 1, \end{cases}$$

we have

$$\mathcal{A}_\tau^{\frac{\alpha}{d}}\left(L_p, X(\{\psi^{(1)},\dots,\psi^{(n)}\})\right) = \dot{B}^\alpha,$$

where $\frac{1}{\tau} = \frac{\alpha}{d} + \frac{1}{p}$.

It should also be mentioned that the results in [BGN04] still hold for noncompactly supported dual wavelets satisfying certain decay conditions. However, then the range of admissible α is smaller, see [BGN04] for details.

Remark 6.1.5. In comparison to the biorthogonal results in Theorem 6.1.3, the bi-frame setting requires stronger assumptions. First, we compare the assumptions on the primal refinable function. For bi-frames, it is supposed to be strongly differentiable and its integer shifts are linearly independent on the unit cube. The biorthogonal results in Theorem 6.1.3 only require weak differentiability, and the linear independence assumption is unnecessary. In fact, a weaker form of linear independence is implicitly provided by the biorthogonality relations, see Remark 5.2.4.

Second, bi-frames require s vanishing moments. For the biorthogonal result in Theorem 6.1.3, we do not explicitly assume any vanishing moments. Nevertheless, since $\varphi \in W^s(L_\infty(\mathbb{R}^d))$ is compactly supported, it is also contained in $W^s(L_1(\mathbb{R}^d))$ and Theorem 1.2.7 yields the reproduction of polynomials up to order $s+1$. Then according to the biorthogonality relations, dual wavelets even have $s+1$ vanishing moments. Since the bi-frame setting does not provide any biorthogonality relations, Theorem 6.1.4 requires the assumption of s vanishing moments.

Third, for small s, i.e, $\frac{s}{d} + \frac{1}{p} \leq 1$, the bi-frame setting yields further restrictions on the range of admissible α. They arise from the localization method, see [BGN04] as well as Subsection 5.3.1.

Theorem 6.1.4 is restricted to dyadic dilation. The results of the present chapter provide an extension to idempotent dilation matrices, see Subsection 1.1.2. Moreover, we derive that, as in the biorthogonal setting, weak differentiability assumptions already suffice.

6.1.2 To Do List

In this subsection, we describe a general procedure to characterize the approximation class. Let X be some Banach space and let Y be a normed, quasi-normed, or even quasi-semi-normed linear space, which is continuously embedded in X. In the sequel, we define some intermediate spaces by the so-called real method of interpolation. First, for $f \in X$ and $t > 0$, let

$$K(f,t,X,Y) := \inf_{g \in Y} \|f-g\|_X + t|g|_Y$$

denote the *Peetre K-functional.* It measures the distance between f and Y with a penalty term depending on t. Then for $0 < \theta < 1$, $0 < q \leq \infty$, the *(real)interpolation space* $[X,Y]_{\theta,q}$ consists of all functions $f \in X$ such that the semi-norm

$$|f|_{[X,Y]_{\theta,q}} := \begin{cases} \left(\int_0^\infty \left(t^{-\theta}K(f,t,X,Y)\right)^q \frac{dt}{t}\right)^{\frac{1}{q}}, & \text{for } 0 < q < \infty, \\ \sup_{0<t<\infty}(t^{\theta}K(f,t,X,Y)), & \text{for } q = \infty, \end{cases} \tag{6.6}$$

is finite. It is equipped with the norm

$$\|f\|_{[X,Y])_{\theta,q}} := \|f\|_X + |f|_{[X,Y]_{\theta,q}},$$

and the semi-norm can be substituted by an equivalent discretization

$$|f|_{[X,Y]_{\theta,q}} \sim \left(\sum_{j\in\mathbb{Z}} \left(2^{\theta j}K(f,2^{-j},X,Y)_{L_p}\right)^q\right)^{\frac{1}{q}}, \tag{6.7}$$

see Chapter 7 in [DL93] as well as the survey article [DeV98]. Moreover, similar to the Lorentz spaces, the number 2 can be replaced by any fixed integer greater than 1.

The definition of interpolation spaces yields the continuous embeddings

$$Y \hookrightarrow [X,Y]_{\theta,q} \hookrightarrow X.$$

Hence, the interpolation space is an intermediate space.

The real method of interpolation is a well-studied topic, and in many situation, the interpolation space is known to be a classical function space. Given $1 < p < \infty$, let s_0, $s_1 \in \mathbb{N}_0$ with $s_0 < s_1$. Then, for $1 \leq q \leq \infty$, $s_0 < \alpha < s_1$, we have

$$\left[W^{s_0}(L_p(\mathbb{R}^d)), W^{s_1}(L_p(\mathbb{R}^d))\right]_{\frac{\alpha-s_0}{s_1-s_0},q} = B_q^{\alpha}(L_p(\mathbb{R}^d)), \tag{6.8}$$

see [DL93, Tri92]. In other words, Besov spaces refine the scale of Sobolev spaces. For $s_0 = 0$, (6.8) reduces to

$$\left[L_p(\mathbb{R}^d), W^s(L_p(\mathbb{R}^d))\right]_{\frac{\alpha}{s},q} = B_q^{\alpha}(L_p(\mathbb{R}^d)),$$

for all $0 < \alpha < s \in \mathbb{N}$. At least for a certain parameter, the interpolation of specific Besov spaces yields again a Besov space: given $1 < p < \infty$, $0 < s_0, s_1 < \infty$, let $s_0 < \alpha < s_1$, then

$$[B^{s_0}, B^{s_1}]_{\frac{\alpha-s_0}{s_1-s_0},\tau} = B^{\alpha}, \tag{6.9}$$

where $\frac{1}{\tau} = \frac{\alpha}{d} + \frac{1}{p}$, see [DL93, Tri92]. According to Chapter 11 in [Pee76], this identity also holds with respect to the homogeneous Besov spaces. Moreover, the interpolation between $L_p(\mathbb{R}^d)$ and $\dot{B}^s$ provides again a homogeneous Besov space, i.e., for $1 < p < \infty$,

$$\left[L_p(\mathbb{R}^d), \dot{B}^s\right]_{\frac{\alpha}{s},\tau} = \dot{B}^\alpha, \tag{6.10}$$

where $\frac{1}{\tau} = \frac{\alpha}{d} + \frac{1}{p}$, cf. [Kyr01], see also [DJP92, CDH00] and the survey article [DeV98].

The real method of interpolation is an important tool for the characterization of the approximation class. The following so-called Jackson and Bernstein estimates provide the connection between approximation and interpolation, cf. [DeV98] and Chapter 7 in [DL93]:

Theorem 6.1.6. *Given $0 < s < \infty$ and a Banach space X, let Y be continuously embedded in X.*

(a) *If the* Jackson inequality

$$\sigma_N(f, \mathcal{D})_X \lesssim N^{-s}|f|_Y, \quad \textit{for all } f \in Y,\ N \in \mathbb{N}, \tag{6.11}$$

holds, then the continuous embedding $[X, Y]_{\frac{\alpha}{s},q} \hookrightarrow \mathcal{A}^\alpha_q(X, \mathcal{D})$ holds, for all $0 < \alpha < s$ and $0 < q \leq \infty$.

(b) *If the* Bernstein inequality

$$|f|_Y \lesssim N^s \|f\|_X, \quad \textit{for all } f \in \Sigma_N(\mathcal{D}),\ N \in \mathbb{N},$$

holds, then the reverse embedding $[X, Y]_{\frac{\alpha}{s},q} \hookleftarrow \mathcal{A}^\alpha_q(X, \mathcal{D})$ holds, for $0 < \alpha < s$ and $0 < q \leq \infty$.

Theorem 6.1.6 is a first step towards the characterization of approximation spaces. By establishing matching Jackson and Bernstein estimates, the approximation class equals an interpolation space. In a next step, one still has to describe the interpolation class by classical function spaces.

We consider best N-term approximation of a wavelet bi-frame. The error is measured in $L_p(\mathbb{R}^d)$, i.e, we require the characterization of the approximation class

$$\mathcal{A}^s_\tau\left(L_p(\mathbb{R}^d), X(\{\psi^{(1)}, \ldots, \psi^{(n)}\})\right).$$

By following the above procedure, in the next section, we establish matching Jackson and Bernstein inequalities with respect to the homogeneous Besov space $\dot{B}^s$. This provides

$$\mathcal{A}^{\frac{\alpha}{d}}_\tau\left(L_p(\mathbb{R}^d), X(\{\psi^{(1)}, \ldots, \psi^{(n)}\})\right) = \left[L_p(\mathbb{R}^d), \dot{B}^s\right]_{\frac{\alpha}{s},\tau}.$$

Then, for $\frac{1}{\tau} = \frac{\alpha}{d} + \frac{1}{p}$, according to (6.10), the right-hand side equals the Besov space $\dot{B}^\alpha$.

6.2 The Characterization of the Approximation Classes

6.2.1 Jackson Estimates

In order to establish a Jackson inequality for wavelet bi-frames with isotropic scalings, we recall a very general Jackson estimate for $\ell_{p,1}$-hilbertian dictionaries in Banach spaces. Then we apply this general result to wavelet bi-frames in $L_p(\mathbb{R}^d)$.

Let $\mathcal{D} = \{g_\kappa : \kappa \in \mathcal{K}\}$ be a dictionary in some Banach space X. For $0 < \tau < \infty$, $0 < q \leq \infty$, and $0 < r < \infty$, denote by $\mathscr{K}_{\tau,q}(X, \mathcal{D}, r)$ the collection

$$\operatorname{clos}_X \left\{ f \in X : f = \sum_{\kappa \in \mathcal{K}'} c_\kappa g_\kappa,\ \mathcal{K}' \subset \mathcal{K},\ \operatorname{card}(\mathcal{K}') < \infty,\ \|(c_\kappa)_{\kappa \in \mathcal{K}'}\|_{\ell_{\tau,q}} \leq r \right\}.$$

Then according to [DT96], the *sparsity class* is defined by

$$\mathscr{K}_{\tau,q}(X, \mathcal{D}) := \bigcup_{r>0} \mathscr{K}_{\tau,q}(X, \mathcal{D}, r),$$

and it is quasi-semi-normed by

$$|f|_{\mathscr{K}_{\tau,q}} := \inf \{ r > 0 : f \in \mathscr{K}_{\tau,q}(X, \mathcal{D}, r) \}.$$

Given $1 < p < \infty$, let $\mathcal{D} = \{g_\kappa : \kappa \in \mathcal{K}\}$ be an $\ell_{p,1}$-hilbertian dictionary. Then the sparsity class has a simpler representation: according to [BGN04, GN04], for $0 < \tau < p$,

$$\mathscr{K}_{\tau,\tau}(X, \mathcal{D}) = \Big\{ \sum_{\kappa \in \mathcal{K}} c_\kappa g_\kappa \in X : (c_\kappa)_{\kappa \in \mathcal{K}} \in \ell_\tau(\mathcal{K}) \Big\} \tag{6.12}$$

and

$$|f|_{\mathscr{K}_{\tau,\tau}} \sim \inf \Big\{ \|(c_\kappa)_{\kappa \in \mathcal{K}}\|_{\ell_\tau} : f = \sum_{\kappa \in \mathcal{K}} c_\kappa g_\kappa \Big\}, \tag{6.13}$$

where the infimum on the right-hand side is actually a minimum.

The following theorem from [GN04] provides a Jackson inequality with respect to the sparsity class:

Theorem 6.2.1 (General Jackson Estimate). *Given $1 < p < \infty$ and $\mathcal{D}$ be an $\ell_{p,1}$-hilbertian dictionary in a Banach space X, let $0 < \alpha < \infty$, $0 < q \leq \infty$, and $\frac{1}{\tau} = \frac{\alpha}{d} + \frac{1}{p}$, then*

$$\sigma_N(f, \mathcal{D})_X \lesssim N^{-\frac{\alpha}{d}} |f|_{\mathscr{K}_{\tau,q}}, \quad \textit{for all } f \in \mathscr{K}_{\tau,q}(X, \mathcal{D}),\ N \in \mathbb{N}. \tag{6.14}$$

Next, in the setting of wavelet bi-frames with isotropic dilation, it turns out that we may replace the sparsity class in Theorem 6.2.1 by a homogeneous Besov space:

Theorem 6.2.2. *Let M be isotropic and $1 < p \leq \infty$. Given a compactly supported wavelet bi-frame $X(\{\psi^{(1)}, \ldots, \psi^{(n)}\})$, $X(\{\widetilde{\psi}^{(1)}, \ldots, \widetilde{\psi}^{(n)}\})$, let the assumptions of Theorem 5.3.12 hold. If α is in the range*

$$0 < \alpha < \begin{cases} s, & \textit{for } \frac{s}{d} + \frac{1}{p} \geq 1, \\ \frac{s}{p(1-\frac{s}{d})}, & \textit{for } \frac{s}{d} + \frac{1}{p} \leq 1, \end{cases}$$

then

$$\sigma_N(f, X(\{\psi^{(1)}, \ldots, \psi^{(n)}\}))_{L_p} \lesssim N^{-\frac{\alpha}{d}} \|f\|_{\dot{B}^\alpha}, \quad \textit{for all } f \in \dot{B}^\alpha,\ N \in \mathbb{N}. \tag{6.15}$$

Proof. By Corollary 5.3.11, the system

$$\{\psi_\lambda^p : \lambda \in \Lambda\}$$

is $\ell_{p,1}(\Lambda)$-hilbertian. Thus, Theorem 6.2.1 is applicable to the sparsity class

$$\mathscr{K}_{\tau,\tau}\left(L_p(\mathbb{R}^d), \{\psi_\lambda^p : \lambda \in \Lambda\}\right),$$

where $\frac{1}{\tau} = \frac{\alpha}{d} + \frac{1}{p}$. This provides

$$\sigma_N(f, X(\{\psi^{(1)}, \ldots, \psi^{(n)}\}))_{L_p} \lesssim N^{-\frac{\alpha}{d}} |f|_{\mathscr{K}_{\tau,\tau}}.$$

We still have to estimate the sparsity norm. Let $\frac{1}{p} + \frac{1}{p'} = 1$, then according to Theorem 5.3.12, for all $f \in \dot{B}^\alpha$,

$$f = \sum_{\lambda \in \Lambda} \langle f, \widetilde{\psi}_\lambda^{p'} \rangle \psi_\lambda^p$$

holds in $L_p(\mathbb{R}^d)$ and

$$(\langle f, \widetilde{\psi}_\lambda^{p'} \rangle)_{\lambda \in \Lambda} \in \ell_\tau(\Lambda).$$

By applying (6.12) and (6.13), this yields

$$|f|_{\mathscr{K}_{\tau,\tau}} \lesssim \left\| (\langle f, \widetilde{\psi}_\lambda^{p'} \rangle)_{\lambda \in \Lambda} \right\|_{\ell_\tau}.$$

Then the norm equivalence of Theorem 5.3.12 concludes the proof. □

6.2.2 Bernstein Estimates

In this subsection, we establish a Bernstein inequality for wavelet bi-frames. It is based on the generalization of the following dyadic Bernstein inequality from [Jia93]:

Theorem 6.2.3. *Given $M = 2\mathcal{I}_d$ and $1 < p < \infty$, let $\varphi \in W^s(L_\infty(\mathbb{R}^d))$, $s \in \mathbb{N}$, be a compactly supported refinable function with linearly independent integer shifts on $(0,1)^d$. Then, for each $0 < \alpha < s$,*

$$\|f\|_{\dot{B}^\alpha} \lesssim N^{\frac{\alpha}{d}} \|f\|_{L_p(\mathbb{R}^d)}, \quad \textit{for all } f \in \Sigma_N\left(X(\{\varphi\})\right).$$

By following the lines of the proof in [Jia93], one verifies that Theorem 6.2.3 still holds for a dilation matrix $M = h\mathcal{I}_d$, where $h \in \mathbb{N}$. This observation is the key ingredient for the proof of the following corollary. It generalizes the dyadic result in [BGN04] regarding idempotent dilation matrices M (recall from Subsection 1.1.2 that a dilation matrix M is called idempotent if there exist $l, h \in \mathbb{N}$ such that $M^l = h\mathcal{I}_d$).

Corollary 6.2.4. *Given an idempotent dilation matrix M and $1 < p < \infty$, let $\varphi \in W^s(L_\infty(\mathbb{R}^d))$ be a compactly supported refinable function with finitely supported mask and with linearly independent integer shifts on $(0,1)^d$. Moreover, let $\psi^{(1)}, \ldots, \psi^{(n)}$ be wavelets with finitely supported sequences $\left(a_k^{(\mu)}\right)_{k \in \mathbb{Z}^d}$ such that*

$$\psi^{(\mu)}(x) = \sum_{k \in \mathbb{Z}^d} a_k^{(\mu)} \varphi(Mx - k), \quad \textit{for } \mu = 1, \ldots, n. \tag{6.16}$$

Then, for $0 < \alpha < s$,

$$\|f\|_{\dot{B}^\alpha} \lesssim N^{\frac{\alpha}{d}} \|f\|_{L_p(\mathbb{R}^d)}, \quad \textit{for all } f \in \Sigma_N(X(\{\psi^{(1)}, \ldots, \psi^{(n)}\})). \tag{6.17}$$

Proof. According to (6.16), we have for each $\mu = 1, \dots, n$,

$$\psi^{(\mu)}(M^j x - k') = \sum_{k \in \mathbb{Z}^d} a_k^{(\mu)} \varphi(M^{j+1}x - Mk' - k), \quad \text{for all } j \in \mathbb{Z},\ k' \in \mathbb{Z}^d.$$

Thus, there exists a constant C_1 such that

$$\psi_{j,k}^{(1)}, \dots, \psi_{j,k}^{(n)} \in \Sigma_{C_1}(X(\{\varphi\})). \tag{6.18}$$

This implies

$$\Sigma_N(X(\{\psi^{(1)}, \dots, \psi^{(n)}\})) \subset \Sigma_{C_1 N}(X(\{\varphi\})). \tag{6.19}$$

Let $(a_k)_{k \in \mathbb{Z}^d}$ be the finitely supported mask of φ, and let l and h be contained in $\mathbb{N}$ such that $M^l = h\mathcal{I}_d$. In the sequel, we verify that there exists a uniform constant C_2 such that, for all $j' \in \mathbb{Z}$ and $k' \in \mathbb{Z}^d$

$$\varphi(M^{j'}x - k') \in \Sigma_{C_2}(\{\varphi(h^j x - k) : j \in \mathbb{Z},\ k \in \mathbb{Z}^d\}). \tag{6.20}$$

Note that we can find $u \in \mathbb{Z}$ and $r \in \mathbb{N}$, $r < l$ such that

$$j' + r = lu.$$

Then r-times applying the refinement equation provides

$$\begin{aligned}
\varphi(M^{j'}x - k') &= \sum_{k_1, \dots, k_r} a_{k_1} \cdots a_{k_r} \varphi(M^r M^{j'} x - M^{r-1}k_1 - \dots - Mk_{r-1} - k_r) \\
&= \sum_{k_1, \dots, k_r} a_{k_1} \cdots a_{k_r} \varphi(M^{lu} x - M^{r-1}k_1 - \dots - Mk_{r-1} - k_r).
\end{aligned}$$

According to $M^l = h\mathcal{I}_d$, the last term is contained in

$$\Sigma_{C^r}(\{\varphi(h^j x - k) : j \in \mathbb{Z},\ k \in \mathbb{Z}^d\}),$$

where C denotes the number of nonzero entries of the mask $(a_k)_{k \in \mathbb{Z}^d}$. Since $r < l$, (6.20) holds with $C_2 = C^{l-1}$.

From (6.20), we derive

$$\Sigma_N(X(\{\varphi\})) \subset \Sigma_{C_2 N}(\{\varphi(h^j x - k) : j \in \mathbb{Z},\ k \in \mathbb{Z}^d\}),$$

which provides with (6.19)

$$\Sigma_N(X(\{\psi^{(1)}, \dots, \psi^{(n)}\})) \subset \Sigma_{C_2 C_1 N}(\{\varphi(h^j x - k) : j \in \mathbb{Z},\ k \in \mathbb{Z}^d\}).$$

Then applying Theorem 6.2.3 to $h\mathcal{I}_d$ yields

$$\begin{aligned}
\|f\|_{\dot{B}^{\alpha}_{\tau}(L_\tau(\mathbb{R}^d))} &\lesssim (C_2 C_1 N)^{\frac{\alpha}{d}} \|f\|_{L_p(\mathbb{R}^d)} \\
&\lesssim N^{\frac{\alpha}{d}} \|f\|_{L_p(\mathbb{R}^d)},
\end{aligned}$$

for all $f \in \Sigma_N(X(\{\psi^{(1)}, \dots, \psi^{(n)}\}))$. □

Remark 6.2.5. The Bernstein inequality of Corollary 6.2.4 is restricted to idempotent dilation matrices. Fortunately, many isotropic dilation matrices addressed in the literature satisfy this additional requirement. Moreover, all our wavelet bi-frames in Chapters 3 and 4 are scaled by idempotent matrices.

Second, we address the arising constants. According to the localization technique, they are already far from being optimal in the Jackson inequality of Theorem 6.2.2, see also Remark 5.3.13. However, our proof of the Bernstein inequality yields to a certain extent an explosion since the constants linearly depend on the number of nonzero entries of the underlying masks and they even exponentially depend on the idempotence of the scaling. Nevertheless, we could derive the qualitative result, and we are convinced that the true constants for the Bernstein inequality are much better than those we used in its proof.

The Bernstein inequality in Corollary 6.2.4 requires that the shifts of the underlying refinable function are linearly independent on the unit cube. Jia conjectures in [Jia93] that the assumption can be removed. However, there is no proof so far, and the application of the corollary requires the verification of this condition. In Section 6.4, we examine the underlying refinable functions of our bi-frames in Chapters 3 and 4 with respect to the linear independence. It turns out that it holds for all examples except for those of the quincunx matrix. Nevertheless, we do not worry since the Bernstein estimate is only a negative statement. The positive Jackson estimate holds without any linear independence assumptions.

6.2.3 Approximation Classes as Besov Spaces

By collecting the results of the previous subsections, it turns out that approximation classes of wavelet bi-frames are essentially Besov spaces. If the assumptions of Theorem 6.2.2 and Corollary 6.2.4 are satisfied, then Theorem 6.1.6 provides

$$\mathcal{A}_\tau^{\frac{\alpha}{d}}\left(L_p(\mathbb{R}^d), X(\{\psi^{(1)},\dots,\psi^{(n)}\})\right) = \left[L_p(\mathbb{R}^d), \dot{B}^s\right]_{\frac{\alpha}{s},\tau}.$$

As already mentioned in Subsection 6.1.2, for $\frac{1}{\tau} = \frac{\alpha}{d} + \frac{1}{p}$, the right-hand side equals the Besov space $\dot{B}^\alpha$.

The following theorem explicitly establishes the final result.

Theorem 6.2.6. *Given an idempotent dilation matrix M and $1 < p < \infty$. Let*

$$X(\{\psi^{(1)},\dots,\psi^{(n)}\}), \quad X(\{\widetilde{\psi}^{(1)},\dots,\widetilde{\psi}^{(n)}\})$$

be a compactly supported wavelet bi-frame. Suppose, that the assumptions of Theorem 5.3.12 *hold. Moreover, let their primal refinable function $\varphi \in W^s(L_\infty(\mathbb{R}^d))$, $s \in \mathbb{N}$, have linearly independent integer shifts on $(0,1)^d$. Then for the range*

$$0 < \alpha < \begin{cases} s, & \text{for } \frac{s}{d} + \frac{1}{p} \geq 1, \\ \frac{s}{p\left(1-\frac{s}{d}\right)}, & \text{for } \frac{s}{d} + \frac{1}{p} \leq 1, \end{cases}$$

we have

$$\mathcal{A}_\tau^{\frac{\alpha}{d}}\left(L_p(\mathbb{R}^d), X(\{\psi^{(1)},\dots,\psi^{(n)}\})\right) = \dot{B}^\alpha, \tag{6.21}$$

where $\frac{1}{\tau} = \frac{\alpha}{d} + \frac{1}{p}$.

In order to apply Theorem 6.2.6 to the wavelet bi-frames constructed in Chapters 3 and 4, we have to verify that their underlying refinable functions have linearly independent integer shifts on the unit cube. Since it turns out that it is not so easy and it requires some effort, we address this topic in the final Section 6.4 of the present chapter. Before, we complete the theoretical framework, and we derive a realization of the best N-term approximation rate in the following Section 6.3.

6.3 N-Term Approximation by Thresholding

Theorem 6.2.6 describes best N-term approximation. In order to implement practical algorithms, we still need a rule for the selection of N particular terms. In other words, we want to realize the best N-term approximation rate. For pairs of biorthogonal wavelet bases, one can simply select the N largest coefficients of the series expansion, cf. [DeV98, Lin05]. This procedure also works for dyadic wavelet bi-frames by thresholding the bi-frame expansion, see [BGN04] for details.

In the following, we extend these results to wavelet bi-frames with general isotropic scalings. Moreover, we allow for more general thresholding operators. They are considered in [BN] with respect to unconditional bases. An analysis of the proof yields that the results hold true for wavelet bi-frames as well. The critical ingredients are the following:

- $\{\psi^p_\lambda : \lambda \in \Lambda\}$ is $\ell_{p,1}$-hilbertian,
- $\left\| \left(\langle f, \widetilde{\psi}^{p'}_\lambda \rangle\right)_{\lambda\in\Lambda} \right\|_{\ell_\tau} \lesssim \|f\|_{\dot{B}^\alpha}$, for all $f \in \dot{B}^\alpha$ and $\frac{1}{\tau} = \frac{\alpha}{d} + \frac{1}{p}$.

Given a wavelet bi-frame $X(\{\psi^{(1)}, \dots \psi^{(n)}\})$, $X(\{\widetilde{\psi}^{(1)}, \dots \widetilde{\psi}^{(n)}\}$ satisfying the assumptions of Theorem 6.2.2, these points are satisfied. Then following [BN], we call a function $\varrho : \mathbb{C} \times \mathbb{R}_+ \to \mathbb{C}$ a *thresholding rule* if

$$|x - \varrho(x, \delta)| \lesssim \min(|x|, \delta)$$

and $|x| \lesssim \delta$ implies $\varrho(x, \delta) = 0$. Given a thresholding rule ϱ, then

$$T_\varrho : \dot{B}^\alpha \times \mathbb{R}_+ \to L_p(\mathbb{R}^d), \quad (f, \delta) \mapsto \sum_{\lambda\in\Lambda} \varrho(\langle f, \widetilde{\psi}^{p'}_\lambda \rangle, \delta)\psi^p_\lambda \tag{6.22}$$

is called the *thresholding operator*. Since $\left(\langle f, \widetilde{\psi}^{p'}_\lambda \rangle\right)_{\lambda\in\Lambda}$ is contained in $\ell_\tau(\Lambda)$, the series (6.22) is actually a finite sum. Note that the operator is applied to the bi-frame coefficients, and one does not allow for thresholding an arbitrary expansion. By denoting

$$N_{f,\delta} := \operatorname{card}\{\lambda \in \Lambda : \varrho(\langle f, \widetilde{\psi}^{p'}_\lambda \rangle, \delta) \neq 0\},$$

thresholding operators realize best N-term approximation:

Theorem 6.3.1. *Under the notation and the assumptions of Theorem 5.3.12, let ϱ be a thresholding rule. Then we have, for α in the range*

$$0 < \alpha < \begin{cases} s, & \text{for } \frac{s}{d} + \frac{1}{p} \geq 1, \\ \frac{s}{p(1-\frac{s}{d})}, & \text{for } \frac{s}{d} + \frac{1}{p} \leq 1, \end{cases}$$

and for all $f \in \dot{B}^{\alpha}$,

$$\|f - T_{\varrho}(f,\delta)\|_{L_p} \lesssim N_{f,\delta}^{-\frac{\alpha}{d}} \|f\|_{\dot{B}^{\alpha}}.$$

According to Theorem 6.3.1, the best N-term approximation rate as described in Theorem 6.2.6 can be realized by thresholding the wavelet bi-frame expansion. Note that Theorem 6.3.1 does not require any linear independence. Hence, even if the assumptions of the Bernstein inequality are not satisfied and so the best N-term approximation is not completely described, thresholding still provides the same approximation rate as predicted by the Jackson inequality.

In the sequel, we consider some simple thresholding rules, see Figure 6.1 for their visualizations. Then, we apply Theorem 6.3.1 to our wavelet bi-frames in Chapters 3 and 4. First, we recall hard- and soft-thresholding. For instance, these rules were already been successfully applied to wavelet expansions in [CDLL98, Tao96].

Example 6.3.2. (a) The rule

$$\varrho_h(x,\delta) = \begin{cases} x, & \text{for } |x| > \delta, \\ 0, & \text{otherwise}, \end{cases}$$

is called *hard-thresholding*.

(b) *Soft-thresholding* is given by

$$\varrho_s(x,\delta) = \begin{cases} x - \frac{x}{|x|}\delta, & \text{for } |x| > \delta, \\ 0, & \text{otherwise}. \end{cases}$$

Hard-thresholding yields discontinuities, and soft-thresholding shrinks all coefficients. The following rule essentially avoids such drawbacks, see [Gao98]:

(c) The thresholding rule

$$\varrho_g(x,\delta) = \begin{cases} x - \frac{\delta^2}{x}, & \text{for } |x| > \delta, \\ 0, & \text{otherwise}, \end{cases}$$

is called *garotte-thresholding*. It is continuous, and large coefficients nearly remain unaltered.

Next, we consider N-term approximation by the wavelet bi-frames in Chapters 3 and 4.

Example 6.3.3. Given a thresholding rule ϱ and $1 < p < \infty$, we address the wavelet bi-frame Laplace (3-2) or Laplace $(3\text{-}2)_{\mathrm{R}}$. According to Theorem 6.3.1, the thresholding operator provides

$$\|f - T_{\varrho}(f,\delta)\|_{L_p} \lesssim N_{f,\delta}^{-\frac{\alpha}{2}} \|f\|_{\dot{B}^{\alpha}},$$

for all $0 < \alpha < 3$.

Thresholding is also applicable to our Checkerboard wavelet bi-frames in Section 3.3.1 as well as to their reduced counterparts in Example 4.3.1. They provide N-term approximation in arbitrary dimensions with an arbitrarily high approximation rate:

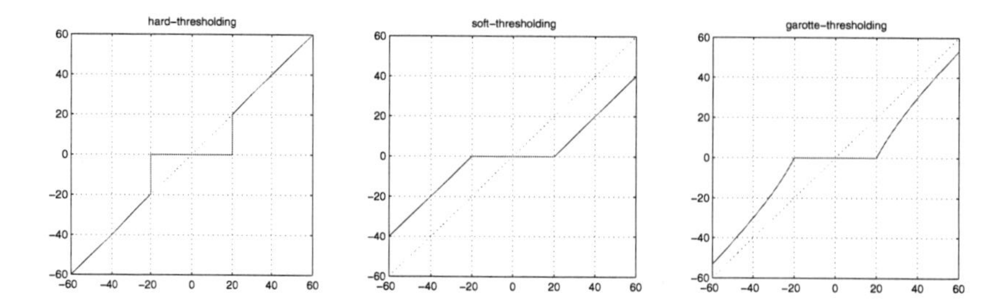

Figure 6.1: Thresholding rules ϱ_h, ϱ_s, and ϱ_g for $\delta = 20$

Example 6.3.4. For arbitrarily large $0 < s \in \mathbb{N}$, there exist Checkerboard bi-frames contained in $W^s(L_\infty(\mathbb{R}^d))$ with s vanishing moments. Hence, we have

$$\|f - T_\varrho(f,\delta)\|_{L_p} \lesssim N_{f,\delta}^{-\frac{\alpha}{d}} \|f\|_{\dot{B}^\alpha},$$

for all $0 < \alpha < s$.

6.4 The Linear Independence on the Unit Cube

The Bernstein inequality in Corollary 6.2.4 requires that the underlying primal refinable function of the wavelet bi-frame has linearly independent integer shifts on the unit cube. The present section is dedicated to verifying this property for our wavelet bi-frames in Chapters 3 and 4.

The linear independence of integer shifts on the unit cube implies global stability as introduced in (1.36). In the dyadic univariate setting, the converse also holds, see [Ron99]:

Theorem 6.4.1. *Let φ be a compactly supported univariate continuous refinable function with respect to dyadic dilation. If its integer shifts are globally linearly independent, then they are also linearly independent on* $(0, 1)$.

All our refinable functions in the examples of Chapters 3 and 4 are fundamental. Thus, they have globally linearly independent integer shifts. Unfortunately, there exist multivariate refinable functions with globally linearly independent integer shifts, which are not linearly independent on $(0, 1)^d$. Hence, Theorem 6.4.1 does not extend to arbitrary multivariate refinable functions, and our aim requires a finer analysis of linear independence.

6.4.1 Box Spline Wavelet Bi-Frames

The present subsection is dedicated to verifying the linear independence condition on $(0, 1)^d$ for the underlying refinable functions of the box spline bi-frames in Subsection 3.3.3 and Example 4.3.3.

In the multivariate setting, Theorem 6.4.1 still holds for box splines, cf. [Jia85]. However, our box spline bi-frames are not constructed from pure box splines, but a box spline convolved with some distribution. Thus, the theorem is not applicable, and we have to find a different approach.

Following [Ron99], we reformulate linear indepence on $(0,1)^d$ in terms of masks. Given a symbol a generating a continuous refinable function φ with respect to dyadic dilation, suppose that $(a_k)_{k\in\mathbb{Z}^d}$ is supported on $[0,L]^d$, $L\in\mathbb{N}$. Let

$$K := [0, L-1]^d \cap \mathbb{Z}^d,$$

fix an order for K, and let

$$\Phi(x) := (\varphi(x+k))_{k\in K}, \quad \text{for } x \in [0,1)^d.$$

Moreover, with $\Gamma^*_{2\mathcal{I}_d} = \{0,1\}^d$, we address

$$\mathscr{A}_{\gamma^*} := \left(a_{\gamma^*+2k-k'}\right)_{k,k'\in K}, \quad \gamma^* \in \Gamma^*_{2\mathcal{I}_d}.$$

Then let $\mathcal{V}$ denote the minimal common invariant subspace of the collection $\{\mathscr{A}_{\gamma^*} : \gamma^* \in \Gamma^*_{2\mathcal{I}_d}\}$, which still contains $\Phi(\frac{1}{2},\dots,\frac{1}{2})$. Finally, let $\mathcal{M}$ be a $\mathrm{card}(K)\times\dim(\mathcal{V})$ matrix such that its columns form a basis for $\mathcal{V}$. This notation holds throughout the present section. Then the following theorem from [Ron99] completely characterizes refinable functions' linear independence on $(0,1)^d$ in terms of their masks:

Theorem 6.4.2. *Let a symbol a with $\mathrm{supp}\,(a_k)_{k\in\mathbb{Z}^d} \subset [0,L]^d$ generate a continuous refinable functions φ with respect to dyadic dilation. Then φ has linearly independent integer shifts on $(0,1)^d$ iff the nonzero rows of the matrix $\mathcal{M}$ are linearly independent.*

The application of Theorem 6.4.2 requires the computation of $\Phi\left(\frac{1}{2},\dots,\frac{1}{2}\right)$. Advantageously, the refinement equation provides

$$\Phi\left(\tfrac{1}{2},\dots,\tfrac{1}{2}\right) = \mathscr{A}_{(1,\dots,1)}\Phi(0),$$

cf. [Ron99]. Hence, we merely need to find $\Phi(0)$, which is often already known. For instance, if φ is fundamental and the order of K starts with 0, then $\Phi(0) = (1,0,\dots,0)^\top$.

Theorem 6.4.2 also requires that the refinable mask is supported on $[0,L]^d$. Since the masks of the refinable functions in Chapters 3 and 4 are centered around zero, we shift them. The following elementary result yields that such shifts do not affect the linear independence:

Lemma 6.4.3. *Let M be a dilation matrix, $l\in\mathbb{Z}^d$, and let φ be $(a_k)_{k\in\mathbb{Z}^d}$ refinable. Then the following holds:*

(a) *The function $\varphi(\cdot - l)$ is $(a_{k-Ml+l})_{k\in\mathbb{Z}^d}$ refinable.*

(b) *If φ has linearly independent integer shifts on A, then also $\varphi(\cdot + l)$, $l\in\mathbb{Z}^d$.*

(c) *If φ has linearly independent integer shifts on A, then also on $A+l$, $l\in\mathbb{Z}^d$.*

We skip the proof of Lemma 6.4.3 since it is direct and elementary. With the help of Theorem 6.4.2, we derive the linear independence on $(0,1)^2$ of one of our underlying wavelet bi-frame refinable functions:

Example 6.4.4. Let φ be the underlying refinable function of the wavelet bi-frame Box Spline (1) as well as Box Spline $(1)_{\mathrm{R}}$. It is fundamental and refinable with respect to dyadic dilation. Denote by $(a_k)_{k\in\mathbb{Z}^2}$ its mask. It is supported on $[-3,3]^2$. In order to apply Theorem 6.4.2, we consider the shifted mask $(a_{k-(3,3)^\top})_{k\in\mathbb{Z}^2}$. According to Lemma 6.4.3, this only yields a shift of φ. Hence, $\Phi(0)$ still consists of zeros except for one single entry, which is equal to one. Finally, $\left(a_{k-(3,3)^\top}\right)_{k\in\mathbb{Z}^2}$ is supported on $[0,6]^2$. Then $L=6$ and $\operatorname{card}(K)=36$. The following results are derived by the computer algebra software Maple. First, we calculate $\dim(\mathcal{V})=30$, i.e., $\mathcal{M}$ has 30 linearly independent rows. Then we verify that there are 6 zero rows in $\mathcal{M}$. Since $\mathcal{M}$ has only 36 rows, its nonzero rows are linearly independent. By applying Theorem 6.4.2, φ has linearly independent integer shifts on $(0,1)^2$.

Our smoother box spline examples of Chapter 4 also satisfy the linear independence assumption of Theorem 6.2.3:

Example 6.4.5. Let φ be the underlying refinable function of the wavelet bi-frame Box Spline $(2)_{\mathrm{R}}$. Note that it is fundamental. Similar to Example 6.4.4, we first apply some shift. Then we have $L=14$ and $\operatorname{card}(K)=196$. Using Maple, we obtain $\dim(\mathcal{V})=154$. Moreover, $\mathcal{M}$ has 42 zero rows. Since $\mathcal{M}$ has exactly 196 rows, the nonzero rows are linearly independent. By applying Theorem 6.4.2, φ has linearly independent integer shifts on $(0,1)^2$.

Since Examples 6.4.4 and 6.4.5 provide the linear independence on $(0,1)^2$, we can apply Theorem 6.2.6 to the characterization of our box spline wavelet bi-frames' approximation classes:

Example 6.4.6. Denote $\psi^{(1)},\dots,\psi^{(n)}$ the primal wavelets of the bivariate wavelet bi-frame Box Spline (1), $(1)_{\mathrm{R}}$, or $(2)_{\mathrm{R}}$. Given $1<p<\infty$, then

$$\mathcal{A}_\tau^{\frac{\alpha}{2}}\left(L_p(\mathbb{R}^2), X(\{\psi^{(1)},\dots,\psi^{(n)}\})\right)=\dot{B}^\alpha$$

holds for all $0<\alpha<1$ with respect to Box Spline (1), $(1)_{\mathrm{R}}$, and it holds for all $0<\alpha<3$ with respect to Box Spline $(2)_{\mathrm{R}}$. According to Theorem 6.3.1, the best N-term approximation rate can be realized by an arbitrary thresholding rule.

We did not verify the local linear independence for Box Spline $(3)_{\mathrm{R}}$ and $(4)_{\mathrm{R}}$. Since their refinable functions share the same structure as those in Examples 6.4.4 and 6.4.5, we are convinced that the independence condition holds.

6.4.2 Quincunx Wavelet Bi-Frames

Next, we address the examples for the quincunx dilation matrix M_q in (3.22). At first glance, Theorem 6.4.2 is useless since it only applies to dyadic dilation. With some trick in the following example, we can circumvent this restriction:

Example 6.4.7. Let φ be the refinable function of the Laplace symbol a in (3.23). Note that φ is fundamental. According to the symmetry of the mask $(a_k)_{k\in\mathbb{Z}^2}$, φ is also refinable with respect to the box spline dilation matrix M_b in (1.6), see [Han04] for details. Since $M_b^2=2\mathcal{I}_2$, one easily verifies that φ is also refinable with respect to dyadic dilation and symbol $a(\xi)a(M_b^\top \xi)$. After a suitable shift, we obtain a mask, which is supported on $[0,4]^2$. Then Maple computes $\dim(\mathcal{V}) = 16$. Since $\mathrm{card}(K) = 16$, the rows of $\mathcal{M}$ are linearly independent, and Theorem 6.4.2 provides the linear independence of the integer shifts on $(0,1)^2$.

Unfortunately, the underlying refinable function φ of our wavelet bi-frames Laplace (1-1) and $(1\text{-}1)_\mathrm{R}$ does not satisfy the linear independence assumption:

Example 6.4.8. Let φ be the underlying refinable function of Laplace (1-1) and $(1\text{-}1)_\mathrm{R}$, and let $(a_k)_{k\in\mathbb{Z}^2}$ be the mask. According to its symmetry, φ is also refinable with respect to the dilation matrix M_b. Since $M_b^2 = 2\mathcal{I}_2$, φ is refinable with respect to dyadic dilation with symbol $a(\xi)a(M_b^\top \xi)$. After a suitable shift, we obtain a mask, which is supported on $[0,12]^2$. Then Maple computes $\dim(\mathcal{V}) = 119$ and $\mathcal{M}$ has 12 zero rows. Hence, its nonzero rows are linearly dependent, and φ does not have linearly independent integer shifts on $(0,1)^2$.

Note that our symmetry trick in Examples 6.4.7 and 6.4.8, i.e., the changeover from M_q to $2\mathcal{I}_2$, enlarges the mask sizes. Hence, already for Laplace (2-2), the complexity is enormous. Nevertheless, since all underlying refinable functions of Laplace (N_1-N_2) and $(N_1\text{-}N_2)_\mathrm{R}$ share the same structure, we are convinced that none of them has linearly independent integer shifts on $(0,1)^2$. For the same reasons, we are also convinced that the underlying refinable function of the bi-frame DGM does not satisfy the linear independence assumption required in Corollary 6.2.4.

6.4.3 Wavelet Bi-Frames in Arbitrary Dimensions

In order to verify the linear independence on $(0,1)^d$ for our Checkerboard wavelet bi-frames in Subsection 3.3.1 and in Example 4.3.1, we require the introduction of another version of linear independence:

Definition 6.4.9. We say a compactly supported distribution φ has *locally linearly independent integer shifts* if its integer shifts are linearly independent on each open subset in $\mathbb{R}^d$.

According to [Ron99], Theorem 6.4.1 extends to local linear independence:

Theorem 6.4.10. *Let φ be a univariate, continuous, dyadic refinable function with compact support. If its integer shifts are globally linearly independent, then they are also locally linearly independent.*

We proceed as follows: we verify that local linear independence of integer shifts is invariant under tensor products of univariate refinable functions. Then according to Theorem 6.4.10, the global linear independence provides that the tensor product has locally linearly independent integer shifts. Finally, we verify that the local linear independence of the tensor product carries over to the underlying refinable functions of the Checkerboard bi-frames.

Proposition 6.4.11. *Let φ_0 be a univariate, continuous, dyadic refinable function with compact support. If its integer shifts are globally linearly independent, then the tensor product $\varphi = \bigotimes_{i=1}^d \varphi_0$ has locally linearly independent integer shifts.*

The following proof of Proposition 6.4.11 is direct. In Appendix A.2, we present an alternative proof in terms of the mask of the refinable function.

Proof. Given a nonempty open subset A in $\mathbb{R}^d$, let x be an arbitrary point in A. Then there exists an open cube $U_x \subset A$, whose edges are parallel to the coordinate axis, and x is contained in U_x. Thus, we have open subsets U_{x_i} in $\mathbb{R}$, $i = 1, \dots, d$, such that

$$U_x = U_{x_1} \times \cdots \times U_{x_d}.$$

According to Theorem 6.4.10, φ_0 has locally linearly independent integer shifts. Hence, for each $i = 1, \dots, d$, the collection

$$B_i := \{\varphi_0(\cdot - k_i) : \operatorname{supp}(\varphi_0(\cdot - k_i)) \cap U_{x_i} \neq \emptyset,\ k_i \in \mathbb{Z}\}$$

is linearly independent. Since linear independence is invariant under tensor products, the collection

$$\begin{aligned} B_1 \otimes \cdots \otimes B_d &= \{\varphi(\cdot - k) : \operatorname{supp}(\varphi_0(\cdot - k_i)) \cap U_{x_i} \neq \emptyset,\ k_i \in \mathbb{Z},\ i = 1, \dots, d\} \\ &= \left\{\varphi(\cdot - k) : \operatorname{supp}(\varphi(\cdot - k)) \cap U_x \neq \emptyset,\ k \in \mathbb{Z}^d\right\} \end{aligned}$$

is also linearly independent. Thus, φ has linearly independent integer shifts on U_x. Since $A = \bigcup_{x \in A} U_x$, the integer shifts of φ are linearly independent on A. This concludes the proof. □

Lemma 6.4.12. *Let $\varphi : \mathbb{R}^{d_1} \to \mathbb{C}$ have locally linearly independent integer shifts, and let $D \in \mathbb{Z}^{d_1 \times d_2}$ be an integer matrix of rank d_1. Then $\varphi(D\cdot) : \mathbb{R}^{d_2} \to \mathbb{C}$ has locally linearly independent integer shifts.*

Proof. Given some nonempty open subset A in $\mathbb{R}^{d_2}$, let

$$\sum_{k \in \mathbb{Z}^{d_2}} c_k \varphi(D(\cdot - k)) = 0, \quad \text{on } A.$$

This implies

$$\sum_{k \in \mathbb{Z}^{d_2}} c_k \varphi(\cdot - Dk) = 0, \quad \text{on } DA.$$

A trivial zero extension yields

$$\sum_{k \in \mathbb{Z}^{d_2}} c_k \varphi(\cdot - Dk) + \sum_{k \in \mathbb{Z}^{d_1} \setminus D\mathbb{Z}^{d_2}} 0 \cdot \varphi(\cdot - k) = 0, \quad \text{on } DA.$$

Since $D : \mathbb{R}^{d_2} \to \mathbb{R}^{d_1}$ is linear and onto, it constitutes an open mapping, i.e., DA is an open subset of $\mathbb{R}^{d_1}$. Hence, the local linear independence of φ provides

$$c_k \varphi(\cdot - Dk) = 0, \quad \text{on } DA, \text{ for all } k \in \mathbb{Z}^{d_2}.$$

Finally, this yields

$$c_k \varphi(D(\cdot - k)) = 0, \quad \text{on } A, \text{ for all } k \in \mathbb{Z}^{d_2},$$

which concludes the proof. □

By applying Proposition 6.4.11 and Lemma 6.4.12, we derive the local linear independence for the underlying refinable function of the Checkerboard bi-frames:

Example 6.4.13. Let φ be the underlying refinable function of the d-dimensional Checkerboard bi-frame in Subsection 3.3.1 or its reduced counterpart in Example 4.3.1. Then according to (3.18), φ is given by

$$\varphi(x) = \varphi_0 \otimes \cdots \otimes \varphi_0 \,(Dx)\,,$$

where φ_0 is a univariate fundamental refinable function, and D is a square matrix with ones in the diagonal as well as above, and zeros elsewhere. Since φ_0 is fundamental, it has globally linearly independent integer shifts. Then according to Proposition 6.4.11, $\varphi_0 \otimes \cdots \otimes \varphi_0$ has locally linearly independent integer shifts. Finally, the application of Lemma 6.4.12 yields that φ has locally linearly independent integer shifts.

Example 6.4.13 verifies that Theorem 6.2.6 can be applied to the Checkerboard bi-frames:

Example 6.4.14. Given $1 < p < \infty$ and an arbitrarily large number $0 < s \in \mathbb{N}$, the Checkerboard family of Subsection 3.3.1 provides a wavelet bi-frame of three wavelets with sufficient smoothness and a sufficient number of vanishing moments such that

$$\mathcal{A}_\tau^{\frac{\alpha}{d}}\left(L_p(\mathbb{R}^d), X(\{\psi^{(1)}, \psi^{(2)}, \psi^{(3)}\})\right) = \dot{B}^\alpha, \quad \text{for all } 0 < \alpha < s,$$

see Theorem 6.2.6. Note that the above equality holds in arbitrary dimensions with only three wavelets. Moreover, for its reduced counterpart of Example 4.3.1, we even derive such an equality with only two wavelets. According to Example 6.3.4, the best N-term approximation rate can be realized by an arbitrary thresholding operator.

Chapter 7

Removing Noise by Solving Variational Problems

In recent years, variational approaches became a valuable tool in signal and image processing concerning compression, noise removal or segmentation, cf. [CDLL98, ROF92]. They are also successfully applied to many other fields of applied mathematics, such as the treatment of operator equations and inverse problems, cf. [DDD04] and references therein.

In the present chapter, we consider variational image and signal denoising. Then given some noisy measurement f, the variational problem consists of finding an approximation g of f, which minimizes the sum of a distance measure and some penalty term. The last-mentioned penalty term usually depends on a so-called regularization parameter, which determines the amount of noise removal, and a careful choice is essential for the success of the method. According to Chambolle, DeVore, Lee, and Lucier in [CDLL98], penalty terms involving a Besov norm provide good results in noise removal from images, but see also [ROF92] for different choices.

In order to derive solutions from practical algorithms, one has to discretize the variational problem. In [CDLL98], the characterization of Besov spaces by biorthogonal wavelet bases reduces the original problem to a discrete variational problem in terms of wavelet coefficients. Then the discrete problem can be explicitly solved, which provides an approximate minimizer of the original one.

One still requires a method for the choice of the regularization parameter, in order to determine a convenient amount of noise removal. Montefusco and Papi proposed the so-called H-curve criterion in [MP03]. It seems a promising approach since it does not need any a-priori knowledge of the noise, and the criterion already provided good results for variational image denoising with respect to the discretization by an orthonormal wavelet basis.

This chapter is dedicated to verifying the potential of multivariate wavelet bi-frames for applicational purposes. We attempt to derive an approximate solution of variational problems with the help of wavelet bi-frames instead of biorthogonal bases, and we hope to verify that the H-curve criterion provides satisfactory results for bi-frames as well.

At first, we establish the variational problem under consideration, and we introduce the H-curve method. Then we recall the discretization by biorthogonal wavelet bases. Since we characterized Besov spaces by means of wavelet bi-frames in Chapter 5, we can also formulate the original variational problem in terms of bi-frame coefficients. As the bases approach in [CDLL98], the solution of the discrete problem provides an approximate minimizer of the original one.

In our numerical experiments of noise removal from images, we discretize the original problem with respect to the bi-frame Laplace (2-2), and we choose the regularization parameter according to the H-curve criterion. We consider additive white noise with different intensities as well as salt&pepper noise. It turns out that the H-curve method provides better denoised images with respect to the visual perception than the choice of the regularization parameter according to the mean square error minimization. Moreover, we also address multiplicative noise, which is much harder to treat than additive noise since it highly depends on the original image. Then the mean square error minimization already provides very good results. Fortunately, the H-curve method yields close mean square errors and the visual results are almost identical. In conclusion, the H-curve criterion is not restricted to orthonormal bases, it provides promising outcomes for bi-frames as well.

7.1 Variational Image Denoising

7.1.1 The Variational Approach

Given some noisy signal f, the reconstruction of the original unperturbed signal $\bar{f}$ either requires some information on $\bar{f}$ or on the noise, and best results can only be expected if one has both. Nevertheless, in order to be very flexible, we consider large classes of noise variants in the following, and we focus on the a-priori knowledge about the original signal, which we may express in terms of membership in a function space. Certainly, the space highly depends on the application, and, for instance, medical images force different classes than radar data or geological measurements. In the present chapter, we consider images without any further specification. Then since ordinary images usually consist of smooth parts and some edges, the space $BV(\mathbb{R}^2)$ of functions of bounded variation seems a good choice for the collection of image representations, see Appendix A.1 for its precise definition, and [ROF92] for its successful applications to image processing. The bivariate Besov space $\dot{B}^1$, $p = 2$, is known to be very close to $BV(\mathbb{R}^2)$, i.e.,

$$\dot{B}^1 \subset BV(\mathbb{R}^2) \subset \dot{B}^1_\infty(L_1(\mathbb{R}^2)),$$

cf. [CDPX99, Mey01], and it constitutes the collection of images in our present setting.

In signal and image processing, many noise variants, such as different kinds of background noise, lead to additive models, i.e.,

$$f = \bar{f} + \varepsilon_1, \tag{7.1}$$

where ε_1 is noise. Since we suppose that $\bar{f}$ is contained in $\dot{B}^1$, and that ε_1 is outside, noise removal means, we try to solely approximate those parts of f, which are contained in $\dot{B}^1$, while avoiding outer parts.

Contrary to additive noise in (7.1), multiplicative noise is generally much harder to treat since it depends on the signal, i.e.,

$$f = \bar{f} + \varepsilon_2 \bar{f}. \tag{7.2}$$

Then noise in a digital camera finally requires the consideration of the superposition of additive and multiplicative noise, i.e.,

$$f = \bar{f} + \varepsilon_1 + \varepsilon_2 \bar{f}, \tag{7.3}$$

cf. [TFG01]. The variational approach, which we shall explain in the following, is applicable to all three noise models (7.1), (7.2), and (7.3). Hence, it is a very flexible tool.

Let us describe the variational noise removal as discussed in [CDLL98] by Chambolle, DeVore, Lee, and Lucier. Since the approach is not restricted to $\dot{B}^1$, we consider a more general setting with multivariate $\bar{f} \in \dot{B}^s$, where $\frac{1}{\tau} = \frac{s}{d} + \frac{1}{2}$ and $\dot{B}^s = \dot{B}^s_\tau(L_\tau(\mathbb{R}^d))$. If we assume $\varepsilon_1 \in L_2(\mathbb{R}^d)$ and $\varepsilon_2 \in L_\infty(\mathbb{R}^d)$, then f given by (7.3) is contained in $L_2(\mathbb{R}^d)$ since $\dot{B}^s \subset L_2(\mathbb{R}^d)$, cf. (5.20). Then for fixed $\delta > 0$, we address the minimization

$$\min_{g \in \dot{B}^s} \left(\|f - g\|_{L_2}^2 + \delta \|g\|_{\dot{B}^s}^\tau \right). \tag{7.4}$$

The exponents 2 and τ are only inserted for computational convenience, and a minimizer $g^{[\delta]}$ approximates f in $L_2(\mathbb{R}^d)$ such that its norm in $\dot{B}^s$ is not too large. Hence, it may constitute a denoised signal.

The parameter δ controls the emphasis of the penalty term $\|g^{[\delta]}\|_{\dot{B}^s}^\tau$, and it determines the amount of noise removal. We shall discuss its choice in the following Subsection 7.1.2.

Remark 7.1.1. We are kind of sloppy concerning the domain of an image. Naturally, the original image $\bar{f}$ as well as the noisy one are represented by some function on a square or a rectangle. This necessitates an extension of the noisy and the original image, and we require a linear extension operator, which is bounded on the addressed Besov spaces. Such operators are derived in [DDD97] and [CDD00]. See also [Ryc99] for some kind of universal extension operator. Note that the one in [DS93] is not applicable since it is nonlinear. Finally, f in (7.4) is already considered as a certain extension from the rectangle to $\mathbb{R}^2$.

7.1.2 The Choice of the Regularization Parameter

Once found a minimizer $g^{[\delta]}$ of (7.4), we still have to choose a specific δ such that the minimizer provides a good representation of the denoised image. In fact, this is a hard problem. On the one hand, if we choose δ too small, then there remains too much noise in the image. On the other hand, large δ provides oversmoothing and we lose too many details.

In order to choose δ, we apply the so-called *H-curve criterion* as proposed by Montefusco and Papi in [MP03]. Let us explain the main idea. Varying $\delta > 0$ provides a curve

$$\left(\log\left(\|f - g^{[\delta]}\|_{L_2}^2 \right), \log\left(\|g^{[\delta]}\|_{\dot{B}^s}^\tau \right) \right) \tag{7.5}$$

in $\mathbb{R}^2$. Then one claims that the curve is concave on a reasonable range of δ, see Figures 7.2, 7.4, and 7.6 for $d = 2$ and $s = 1$, and one chooses δ_H according to the maximum absolute value of the curvature. This choice may provide a good balance between the penalty term and the error of approximation in $L_2(\mathbb{R}^d)$. Note that the approach does not require any information about the noise. Hence, it is a very flexible tool, and it should be mentioned that is has already been successfully applied to variational noise removal involving orthonormal wavelet bases in [MP03].

7.2 Discretization

In order to derive the minizer of (7.4) by a practical algorithm, we have to discretize the original problem. Such discretizations do usually not provide an exact minimizer, but only an *approximate minimizer* $g^{[\delta]}$ of (7.4), i.e., for all $g \in \dot{B}^s$,

$$\left\|f - g^{[\delta]}\right\|_{L_2}^2 + \delta \left\|g^{[\delta]}\right\|_{\dot{B}^s}^{\tau} \lesssim \|f - g\|_{L_2}^2 + \delta \|g\|_{\dot{B}^s}^{\tau}.$$

Fortunately, it turns out that these approximate solutions of (7.4) are sufficient for applicational purposes, cf. [CDLL98].

7.2.1 Discretization by Biorthogonal Wavelets

In the present subsection, we follow [CDLL98], and we discretize the problem (7.4) by a given pair of biorthogonal wavelet bases

$$X(\{\psi^{(1)}, \ldots, \psi^{(n)}\}), \quad X(\{\widetilde{\psi}^{(1)}, \ldots, \widetilde{\psi}^{(n)}\}) \tag{7.6}$$

characterizing $\dot{B}^s$ as in Theorem 5.2.6. Let

$$\widetilde{F}^* : L_2(\mathbb{R}^2) \to \ell_2(\Lambda), \quad f \mapsto \left(\left\langle f, \widetilde{\psi}_\lambda \right\rangle\right)_{\lambda \in \lambda},$$

be the associated analysis operator, then according to the characterization of $L_2(\mathbb{R}^d)$ and $\dot{B}^s$, the problem (7.4) can be replaced by the discrete minimization problem

$$\min_{w \in \ell_\tau} \left(\left\|\widetilde{F}^* f - w\right\|_{\ell_2}^2 + \delta \|w\|_{\ell_\tau}^{\tau} \right). \tag{7.7}$$

Let $w^{[\delta]}$ be an exact minimizer of (7.7). Since we assume that (7.6) constitute Riesz bases, the operator $\widetilde{F}^*$ is boundedly invertible, and

$$g^{[\delta]} = (\widetilde{F}^*)^{-1} w^{[\delta]} \tag{7.8}$$

provides an approximate minimizer of the original variational problem (7.4), see [CDLL98] for details. Advantageously, the discrete minimization (7.7) is decoupled, and it can be separately minimized for each $\lambda \in \Lambda$, i.e., we minimize

$$\min_{w_\lambda} \left(|v_\lambda - w_\lambda|^2 + \delta |w_\lambda|^\tau \right),$$

where $v := \widetilde{F}^* f$. Thus, we reduced the complicated continuous minimization to a much simpler discrete problem.

According to the results in [CDLL98], the hard-threshold choice

$$w_\lambda = \begin{cases} v_\lambda, & \text{for } |v_\lambda| > \delta^{\frac{1}{2-\tau}}, \\ 0, & \text{otherwise,} \end{cases} \tag{7.9}$$

provides a good approximation of $w^{[\delta]}$ since it minimizes (7.7) within a factor of 4 of the exact minimum. In other words, it provides an approximate minimizer of the discrete problem. For general values of s, one only has an implicit description of the exact minimizer, but as s goes to infinity and so τ goes to zero, the exact minimizer $w^{[\delta]}$ of (7.7) converges to hard-thresholding

$$w_\lambda^{[\delta]} = \begin{cases} v_\lambda, & \text{for } |v_\lambda| > \sqrt{\delta}, \\ 0, & \text{otherwise,} \end{cases}$$

cf. [Lor07]. For the specific case $s = \frac{1}{2}d$, the exact minimizer $w^{[\delta]}$ can be explicitly determined by soft-thresholding

$$w_\lambda^{[\delta]} = \begin{cases} v_\lambda - \frac{v_\lambda}{|v_\lambda|}\frac{\delta}{2}, & \text{for } |v_\lambda| > \frac{\delta}{2}, \\ 0, & \text{otherwise,} \end{cases} \tag{7.10}$$

see [CDLL98].

Recall that hard- and soft-thresholding also arised as a realization of best N-term approximation, cf. Example 6.3.2 in Chapter 6. Hence, variational noise removal can be considered as an N-term approximation of f, where the number of terms determines the amount of noise removal.

Remark 7.2.1. In [DJ94], Donoho and Johnston address the problem of image denoising via some statistical approach. They derive some soft-thresholding algorithm, and an asymptotically optimal threshold parameter is determined. However, their results are restricted to orthonormal bases and additive white noise. In the present chapter, we require more flexibility with respect to the wavelet system, and we also consider different noise models beyond additive white noise, see (7.2) and (7.3). Hence, we address a more far-reaching setting.

We still have to choose δ_H according to the H-curve criterion. The discretization of the curve in (7.5) yields

$$\left(\log\left(\|v - w^{[\delta]}\|_{\ell_2}^2\right), \log\left(\|w^{[\delta]}\|_{\ell_\tau}^\tau\right)\right), \tag{7.11}$$

and then we determine δ_H according to the maximal curvature of this curve.

7.2.2 Discretization by Wavelet Bi-Frames

In the previous Subsection 7.2.1, we discretized the original variational problem with respect to a pair of biorthogonal wavelet bases. In the following, we extend the discretization to wavelet bi-frames.

Let the collection in (7.6) constitute a wavelet bi-frame with respect to isotropic scaling and no longer a pair of biorthogonal bases. Given $s > 0$, let k denote an integer strictly larger than s and $\frac{d}{2}$. We assume that all wavelets of the bi-frame have at least k vanishing moments and that they are contained in $W^k(L_\infty(\mathbb{R}^d))$. Then according to Theorem 5.3.12, we still have the norm equivalences provided that there exists a pair of biorthogonal reference bases. However, the dual analysis operator $\widetilde{F}^*$ is no longer onto, i.e., it is not invertible, and (7.8) makes no sense any more. One has to restrict the minimization to the range of $\widetilde{F}^*$. From a computational point of view, this is problematic.

We circumvent the restriction by the following alternative. First, we do not care about the range of $\widetilde{F}^*$, and we minimize (7.7) on $\ell_\tau(\Lambda)$. Then applying the orthogonal projection onto the range of $\widetilde{F}^*$ seems to be a good compromise between exactness and computational efficiency. The orthogonal projection is provided by

$$P : \ell_2(\Lambda) \to \ell_2(\Lambda), \quad (c_\lambda)_{\lambda\in\Lambda} \mapsto \left(\sum_{\lambda\in\Lambda} \left\langle \widetilde{S}^{-1}\widetilde{\psi}_\lambda, \widetilde{\psi}_{\lambda'} \right\rangle c_\lambda\right)_{\lambda'\in\Lambda}, \tag{7.12}$$

where $\widetilde{S} = \widetilde{F}\widetilde{F}^*$ is the dual frame operator, see Section 5.3 in [Chr03]. This means $P = \widetilde{F}^* F_{\widetilde{S}}$, where $F_{\widetilde{S}}$ is the synthesis operator with respect to the canonical dual frame of $\{\widetilde{\psi}_\lambda : \lambda \in \Lambda\}$.

However, the orthogonal projection (7.12) inherits some drawbacks since it requires the inversion of the frame operator $\widetilde{S}$ and the elements $\widetilde{S}^{-1}\widetilde{\psi}_\lambda$ might not possess the wavelet structure. In order to avoid such invonveniences, let us apply the bi-frame concept once more, and we consequently substitute the primal wavelets ψ_λ for $\widetilde{S}^{-1}\widetilde{\psi}_\lambda$. In other words, the orthogonal projection (7.12) is replaced with the operator $\widetilde{F}^*F$, where

$$F : \ell_2(\Lambda) \to L_2(\mathbb{R}^2), \quad (c_\lambda)_{\lambda\in\Lambda} \mapsto \sum_{\lambda\in\Lambda} c_\lambda \psi_\lambda$$

is the primal synthesis operator. This still provides a nonorthogonal projection onto the range of $\widetilde{F}^*$.

According to Theorem 5.3.1, $\widetilde{F}^*Fw^{[\delta]}$ is contained in $\ell_\tau(\Lambda)$. Since the dual wavelets constitute a frame, $\widetilde{F}^*$ is injective, i.e., it is invertible on its range, and we can define

$$g^{[\delta]} := \left(\widetilde{F}^*_{|\operatorname{range}(\widetilde{F}^*)}\right)^{-1} \widetilde{F}^*Fw^{[\delta]} = Fw^{[\delta]}. \tag{7.13}$$

Remark 7.2.2. Note that $g^{[\delta]} = Fw^{[\delta]}$ in (7.13) finally yields that we do not require any projection.

Since F maps $\ell_\tau(\Lambda)$ into $\dot{B}^s$, the function $g^{[\delta]}$ is indeed contained in $\dot{B}^s$, and in the sequel, we shall verify that it is an approximate minimizer of (7.4). By applying the norm equivalences of Theorem 5.3.12 and since $F\widetilde{F}^*$ equals the identity, we obtain

$$\begin{aligned}\left\|f - g^{[\delta]}\right\|_{L_2}^2 + \delta\left\|g^{[\delta]}\right\|_{\dot{B}^s}^\tau &\lesssim \left\|\widetilde{F}^*f - \widetilde{F}^*g^{[\delta]}\right\|_{\ell_2}^2 + \delta\left\|\widetilde{F}^*g^{[\delta]}\right\|_{\ell_\tau}^\tau \\ &= \left\|\widetilde{F}^*F\widetilde{F}^*f - \widetilde{F}^*Fw^{[\delta]}\right\|_{\ell_2}^2 + \delta\left\|\widetilde{F}^*Fw^{[\delta]}\right\|_{\ell_\tau}^\tau.\end{aligned}$$

Since we have a bi-frame, the operator

$$\widetilde{F}^*F : (c_\lambda)_{\lambda\in\Lambda} \mapsto \left(\sum_{\lambda\in\Lambda} \left\langle \psi_\lambda, \widetilde{\psi}_{\lambda'} \right\rangle c_\lambda\right)_{\lambda'\in\Lambda},$$

is bounded on $\ell_2(\Lambda)$, and, according to Theorem 5.3.1, it is also bounded on $\ell_\tau(\Lambda)$. This yields

$$\left\|f - g^{[\delta]}\right\|_{L_2}^2 + \delta\left\|g^{[\delta]}\right\|_{\dot{B}^s}^\tau \lesssim \left\|\widetilde{F}^*f - w^{[\delta]}\right\|_{\ell_2}^2 + \delta\left\|w^{[\delta]}\right\|_{\ell_\tau}^\tau.$$

By applying that $w^{[\delta]}$ is a minimizer of (7.7), we obtain, for all $g \in \dot{B}^s$,

$$\left\|f - g^{[\delta]}\right\|_{L_2}^2 + \delta\left\|g^{[\delta]}\right\|_{\dot{B}^s}^\tau \lesssim \left\|\widetilde{F}^*f - \widetilde{F}^*g\right\|_{\ell_2}^2 + \delta\left\|\widetilde{F}^*g\right\|_{\ell_\tau}^\tau,$$

and then the norm equivalences of Theorem 5.3.12 lead to

$$\left\|f - g^{[\delta]}\right\|_{L_2}^2 + \delta \left\|g^{[\delta]}\right\|_{\dot{B}^s}^{\tau} \lesssim \|f - g\|_{L_2}^2 + \delta \|g\|_{\dot{B}^s}^{\tau}.$$

Thus, $g^{[\delta]}$ is an approximate minimizer of the original variational problem (7.4).

For $s = \frac{d}{2}$, soft-thresholding yields the exact minimizer of the discret problem. At least in this case, it turns out that each thresholding rule provides an approximate minimizer of the original problem:

Theorem 7.2.3. *Given $s = \frac{d}{2}$, a thresholding rule ϱ, and an isotropic dilation matrix M, let $X(\{\psi^{(1)}, \dots, \psi^{(n)}\})$, $X(\{\widetilde{\psi}^{(1)}, \dots, \widetilde{\psi}^{(n)}\})$ be a compactly supported wavelet bi-frame, which characterizes $\dot{B}^s$. Then the thresholding operator*

$$T_{\varrho}(f, \delta) = \sum_{\lambda \in \Lambda} \varrho(\langle f, \widetilde{\psi}_{\lambda} \rangle, \delta) \psi_{\lambda}$$

provides an approximate minimizer of (7.4).

Proof. We have already verified that an exact discrete minimizer yields an approximate minimizer of the original problem. By following the lines once more, one verifies that an approximate discrete minimizer already suffices. In the sequel, we verify that thresholding provides such an approximate minimizer of the discrete problem.

Since ϱ is a thresholding rule, there are two positive constants C_1 and C_2 such that, for all $x \in \mathbb{C}$ and $\delta > 0$,

$$|x - \varrho(x, \delta)| \leq C_1 \min(|x|, \delta), \tag{7.14}$$

and $|x| \leq C_2 \delta$ implies $\varrho(x, \delta) = 0$. Given the exact discrete minimizer $w^{[\delta]}$, we apply the short-hand notation

$$\begin{aligned} K_{\lambda} &:= \left|v_{\lambda} - w_{\lambda}^{[\delta]}\right|^2 + \delta \left|w_{\lambda}^{[\delta]}\right|, \\ G_{\lambda} &:= |v_{\lambda} - \varrho(v_{\lambda}, \delta)|^2 + \delta \left|\varrho(v_{\lambda}, \delta)\right|. \end{aligned}$$

Note that $w^{[\delta]}$ can be derived by soft-thresholding of $(v_{\lambda})_{\lambda \in \Lambda}$. In order to verify that $G_{\lambda} \lesssim K_{\lambda}$, where the constant is independent on λ, v_{λ}, and δ, we consider five cases. First, for $0 \leq |v_{\lambda}| \leq \min(\frac{\delta}{2}, C_2 \delta)$, we have $G_{\lambda} = K_{\lambda}$.

Second, if $\frac{\delta}{2} \leq |v_{\lambda}| \leq C_2 \delta$, then on the one hand, we obtain

$$G_{\lambda} = |v_{\lambda}|^2 \leq C_2^2 \delta^2.$$

On the other hand, we have

$$K_{\lambda} = \frac{\delta^2}{4} + \delta \left|v_{\lambda} - \frac{\delta}{2}\right|,$$

which yields $G_{\lambda} \leq 4 C_2^2 K_{\lambda}$.

Third, we consider $C_2 \delta \leq |v_{\lambda}| \leq \frac{\delta}{2}$. Since (7.14) also yields $|\varrho(v_{\lambda}, \delta)| \leq (C_1 + 1)|v_{\lambda}|$, we obtain

$$\begin{aligned} G_{\lambda} &\leq C_1^2 \min(|v_{\lambda}|^2, \delta^2) + \delta |\varrho(v_{\lambda}, \delta)| \\ &\leq C_1^2 |v_{\lambda}|^2 + C_2^{-1} |v_{\lambda}| (C_1 + 1) |v_{\lambda}| \\ &\leq (C_1^2 + C_2^{-1} C_1 + C_2^{-1}) |v_{\lambda}|^2 = (C_1^2 + C_2^{-1} C_1 + C_2^{-1}) K_{\lambda}. \end{aligned}$$

Fourth, if $\max(\frac{\delta}{2}, C_2\delta) \le |v_\lambda| \le \delta$, then we have

$$\begin{aligned} G_\lambda &\le C_1^2 \min(|v_\lambda|^2, \delta^2) + \delta|\varrho(v_\lambda, \delta)| \\ &\le C_1^2\delta^2 + \delta(C_1+1)|v_\lambda| \le (C_1^2 + C_1 + 1)\delta^2. \end{aligned}$$

Then applying

$$K_\lambda = \frac{\delta^2}{4} + \delta \left| v_\lambda - \frac{\delta}{2} \right| \ge \frac{\delta^2}{4}$$

yields $G_\lambda \le 4(C_1^2 + C_1 + 1)K_\lambda$.

Fifth, given $\max(\delta, C_2\delta) \le |v_\lambda|$, we obtain

$$\begin{aligned} G_\lambda &\le C_1^2 \min(|v_\lambda|^2, \delta^2) + \delta|\varrho(v_\lambda, \delta)| \\ &\le C_1^2\delta^2 + \delta(C_1+1)|v_\lambda|. \end{aligned}$$

Since the reverse triangle inequality yields

$$K_\lambda = \frac{\delta^2}{4} + \delta \left| v_\lambda - \frac{\delta}{2} \right| \ge \frac{\delta^2}{4} + \delta \left| |v_\lambda| - \frac{\delta}{2} \right| \ge \frac{\delta^2}{4} + \delta|v_\lambda|\frac{1}{2},$$

we have estimated $G_\lambda \le 4\max(C_1^2, \frac{1}{2}(C_1+1))K_\lambda$.

Finally, we established that an arbitrary thresholding rule provides an approximate minimizer of the discrete variational problem. This concludes the proof. □

Concerning the H-curve method, we can also discretize the curve (7.5) by a wavelet bi-frame. Then the discrete version in terms of bi-frame coefficients looks like the one in (7.11).

7.3 Numerical Results

For our experiments, we suppose that the noisy image f is still contained in $L_2(\mathbb{R}^2)$, while the unperturbed image is an element of $\dot{B}^1$. Then we consider the variational problem

$$\min_{g \in \dot{B}^1} \left(\|f - g\|_{L_2}^2 + \delta \|g\|_{\dot{B}^1} \right). \tag{7.15}$$

We discretize (7.15) with respect to the bi-frame Laplace (2-2), which characterizes the bivariate Besov space $\dot{B}^1$, see Example 5.3.16. Then according to Theorem 7.2.3, each thresholding rule provides an approximate minimizer of (7.15). Nevertheless, we derive $g^{[\delta]}$ by applying soft-thresholding since it corresponds to the exact minimizer of the associated discrete problem. Then we determine the threshold parameter δ_H by the H-curve criterion. In order to evaluate its choice, we compare the results with another choice δ_{MSE} according to the minimization of the *mean square error*, i.e., for $X, Y : \{1, \ldots, 512\}^2 \to \mathbb{R}$ representing the pixel values of images of size 512×512, we consider

$$\begin{aligned} \mathrm{MSE}(X,Y) &:= \frac{1}{512^2} \sum_{i,j=1}^{512} |X_{i,j} - Y_{i,j}|^2, \\ \mathrm{RSME}(X,Y) &:= \sqrt{\mathrm{MSE}(X,Y)}, \end{aligned}$$

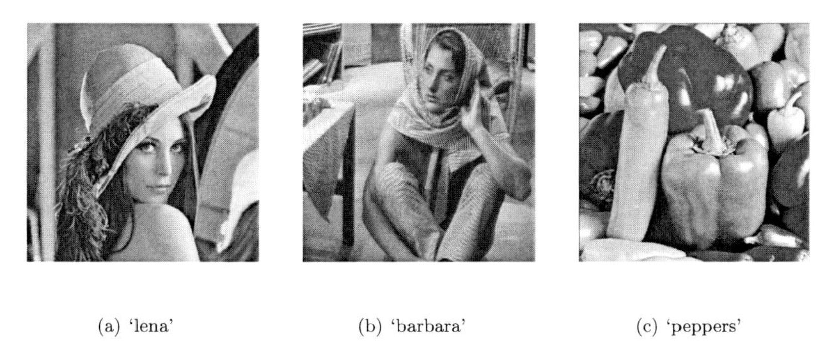

(a) ‘lena’ (b) ‘barbara’ (c) ‘peppers’

Figure 7.1: Original images

and then let δ_{MSE} be the threshold parameter, which corresponds to a minimal MSE. It should be mentioned that the MSE is one of the standard error measures in literature, but such mathematical error measures do often not reflect the visual perception. Finally, our numerical considerations are dedicated to verify that δ_H outperforms δ_{MSE}. In other words, we attempt to verify that the H-curve criterion yields better results with respect to the visual perception than the MSE minimization.

We consider 8-bit grayscale images ‘lena’, ‘barbara’, and ‘peppers’ of size 512×512, see Figure 7.1. Then we corrupt them by different kinds of noise, such as additive and multiplicative noise in (7.1), (7.2), and (7.3). In order to decompose the image almost completely in wavelet coefficients, we calculate 10 scales of the transform. To avoid random anomalies, we average over 20 iterations.

Remark 7.3.1. Since implementations of the extension operators addressed in Remark 7.1.1 are very complicated, we simply apply the periodic extension of an image. This method is very popular in image processing, cf. [Mal99, SN96]. However, it has to be mentioned that it is not admissible by our theory. One may overcome this looseness with the concept by periodic wavelets, so that the Besov characterization can be applied to periodic functions, cf. [Dau92].

7.3.1 Additive Gaussian White Noise

In order to simulate background noise, we shall consider additive Gaussian white noise, which leads to the noise model (7.1), where ε_1 is essentially Gaussian distributed with zero mean, cf. [Mal99]. We address standard deviations $\sigma_1 = 10$, 25, and 40, which we refer to low, medium, and strong noise.

It turns out that, for all three test images, the discrete analog of the curve in (7.5) is concave, and the H-curve criterion is applicable, cf. Figure 7.2 for ‘lena’ and $\sigma_1 = 10$, Figure 7.4 for ‘barbara’ and $\sigma_1 = 25$, and Figure 7.6 for ‘peppers’ and $\sigma_1 = 40$.

At first, let us remove noise from ‘lena’. A comparison of the threshold parameter δ_H of the H-curve method with the MSE minimizer δ_{MSE} as well as their RSMEs is presented in Table 7.1. For low noise, δ_H is larger than δ_{MSE}, but both RMSEs are still quite close. Considering stronger noise such as $\sigma_1 = 25, 40$, the threshold δ_H is smaller than

σ_1	δ_H	$\delta_H/\delta_{\mathrm{MSE}}$	RMSE δ_H	RMSE δ_{MSE}
10	13.61	1.30	6.17	5.94
25	25.40	0.74	11.35	10.62
40	35.72	0.56	17.69	13.78

Table 7.1: 'lena', Gaussian white noise

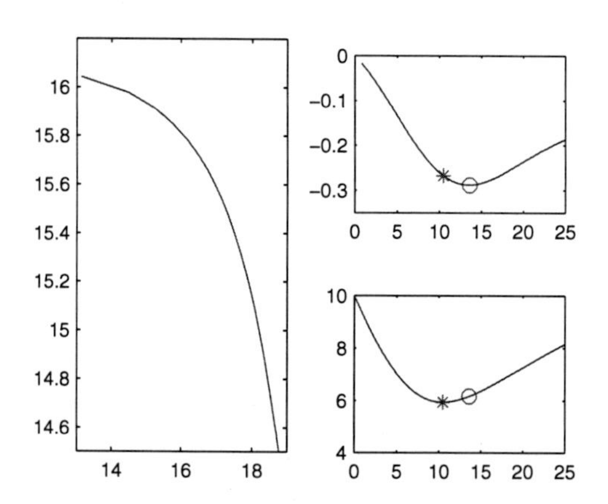

Figure 7.2: 'lena', Gaussian white noise, $\sigma_1 = 10$, left: concave curve, top right: curvature/δ, down left: RMSE/δ with δ_{MSE} ($*$) and δ_H ($\circ$).

σ_1	δ_H	$\delta_H/\delta_{\text{MSE}}$	RMSE δ_H	RMSE δ_{MSE}
10	23.06	1.88	11.33	9.59
25	29.21	0.93	14.69	14.65
40	38.09	0.67	19.93	17.96

Table 7.2: 'barbara', Gaussian white noise

σ_1	δ_H	$\delta_H/\delta_{\text{MSE}}$	RMSE δ_H	RMSE δ_{MSE}
10	13.21	1.28	6.28	6.09
25	26.43	0.78	11.31	10.79
40	36.64	0.59	17.53	14.12

Table 7.3: 'peppers', Gaussian white noise

δ_{MSE}, and we remove less noise than with δ_{MSE}. Nevertheless, we may hope for more details, and this is supported by Figure 7.3, where the noisy image is on the left, in the center is the H-curved denoised image, and we have the optimal choice with respect to MSE minimization on the right. For $\sigma_1 = 25$, the choice δ_H keeps more details than δ_{MSE} and so in the H-curve denoised image remains slightly more noise. Since it is generally more important to preserve details than removing all the noise, one may consider δ_H as the better choice. In case of stronger noise $\sigma_1 = 40$, this effect is more obvious. Then δ_{MSE} almost removes all the noise, but one also loses most of the details. The choice δ_H allows for more noise, but one can still recognize a lot of details. Thus, the H-curve method leads to far better visual results, although its MSE is noticeably larger than the minimum. Especially for stronger noise, the MSE minimization does not seem to reflect the visual perception of a minimal error, and the bi-frame is well suited for the H-curve method.

For low noise $\sigma_1 = 10$, both denoised images are more or less visually identical. Maybe, δ_H provides some more smoothing, but it is hard to recognize.

Next, we address 'barbara' with the same noise levels as before, see Table 7.2 and Figure 7.4 for the numerical results. In order to evaluate the visual results, see Figure 7.5. For low noise, the choice δ_H seems slightly too large, and one might prefer δ_{MSE}. Nevertheless, the visual differences are very small. For $\sigma_1 = 25$, both δ_H and δ_{MSE} are very close, and they provide satisfactory visual results. For strong noise, we definitely prefer δ_H, although there still remains more noise, because we also preserve much more details. In conclusion, 'barbara' supports the good performance of the H-curve method with respect to 'lena', and especially for stronger noise, the choice by the H-curve criterion provides better results with respect to the visual perception than the MSE minimization.

Finally, we address 'peppers', corrupted by additive Gaussian white noise, see Table 7.3 and Figure 7.6 for numerical results. The denoised images are presented in Figure 7.7. For low noise, δ_H and δ_{MSE} yield very similar results. Given $\sigma_1 = 25$, the H-curve criterion is better since it keeps more details than δ_{MSE}. For strong noise, the effect is more noticeable, and δ_H is obviously the better choice.

In conclusion, for all three test images, the H-curve method provides better visual results with respect to additive Gaussian white noise than the MSE minimization.

Figure 7.3: 'lena', Gaussian white noise, from top to down: $\sigma_1 = 10, 25, 40$, from left to right: noisy, δ_H, δ_{MSE}

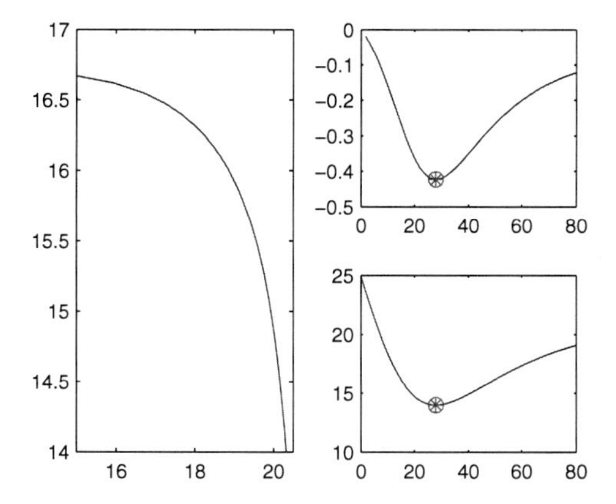

Figure 7.4: 'barbara', Gaussian white noise, $\sigma_1 = 25$, left: concave curve, top right: curvature/δ, down right: RMSE/δ with δ_{MSE} ($*$) and δ_H ($\circ$).

image	δ_H	$\delta_H/\delta_{\text{MSE}}$	RMSE δ_H	RMSE δ_{MSE}
'lena'	51.92	0.42	27.17	20.29
'barbara'	50.58	0.41	29.78	23.93
'peppers'	53.39	0.42	22.64	29.45

Table 7.4: Salt&Pepper noise

7.3.2 Salt&Pepper Noise

In digital cameras, there often occur pixel errors, which can be simulated by salt&pepper noise, i.e., one flips a certain percentage of pixels either to 0 or to 255. We consider a uniformly distributed spatial density of 15%.

The results of all three test images are presented in Figure 7.8 and Table 7.4. For δ_{MSE}, the denoised images look very blurry, and it is obvious that the MSE minimization removes too many details. The threshold δ_H is less than $\frac{1}{2}\delta_{\text{MSE}}$, and it allows for remaining noise. Nevertheless, since the choice δ_H also keeps a lot more details, it definitely provides the best results with respect to the visual perception. In conclusion, for salt&pepper noise, the H-curve method outperforms the MSE minimization.

7.3.3 Multiplicative Gaussian White Noise

Removal of multiplicative noise is generally much harder than removing additive noise since it depends on the signal. We address (7.2), where ε_2 is Gaussian white noise. While one considers the standard deviations 0.1 and 0.2 in [HP05], we choose the average $\sigma_2 = 0.15$, see Figure 7.9 for the denoised images. The numerical results are given in Figure 7.10, where we visualize the differences between δ_H and δ_{MSE} as well as the different RMSEs. It turns out that δ_{MSE} is already a very good choice with respect to the visual perception. Fortunately, the H-curve criterion provides a threshold δ_H, which

Figure 7.5: 'barbara', Gaussian white noise, from top to down: $\sigma_1 = 10, 25, 40$, from left to right: noisy, δ_H, δ_{MSE}

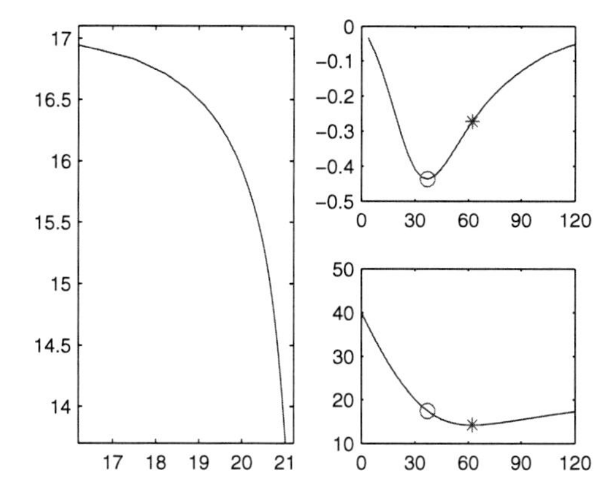

Figure 7.6: 'peppers', Gaussian white noise, $\sigma_1 = 40$, left: concave curve, top right: curvature/δ, down right: RMSE/δ with δ_{MSE} ($*$) and δ_H ($\circ$).

is very close to δ_{MSE}. Hence, their denoised images are visually identical, and δ_H also provides optimal results. Note that Figure 7.10 shows that δ_H and δ_{MSE} almost coincide, and so do their RMSEs.

7.3.4 Additive and Multiplicative Gaussian White Noise

Since the simulation of noise in a digital camera of a so-called CMOS photodiode active pixel sensor (APS) requires the combination of additive and multiplicative white noise, cf. [TFG01], we finally consider (7.3), where ε_1 and ε_2 are Gaussian white noise with standard deviation $\sigma_1 = 20$ and $\sigma_2 = 0.1$, respectively. The denoised images are presented in Figure 7.12, while Figure 7.11 provides a visual comparison between δ_H and δ_{MSE}. The MSE minimization already yields good visual results and δ_H is very close to δ_{MSE}. Hence, the H-curve method provides a satisfactory choice.

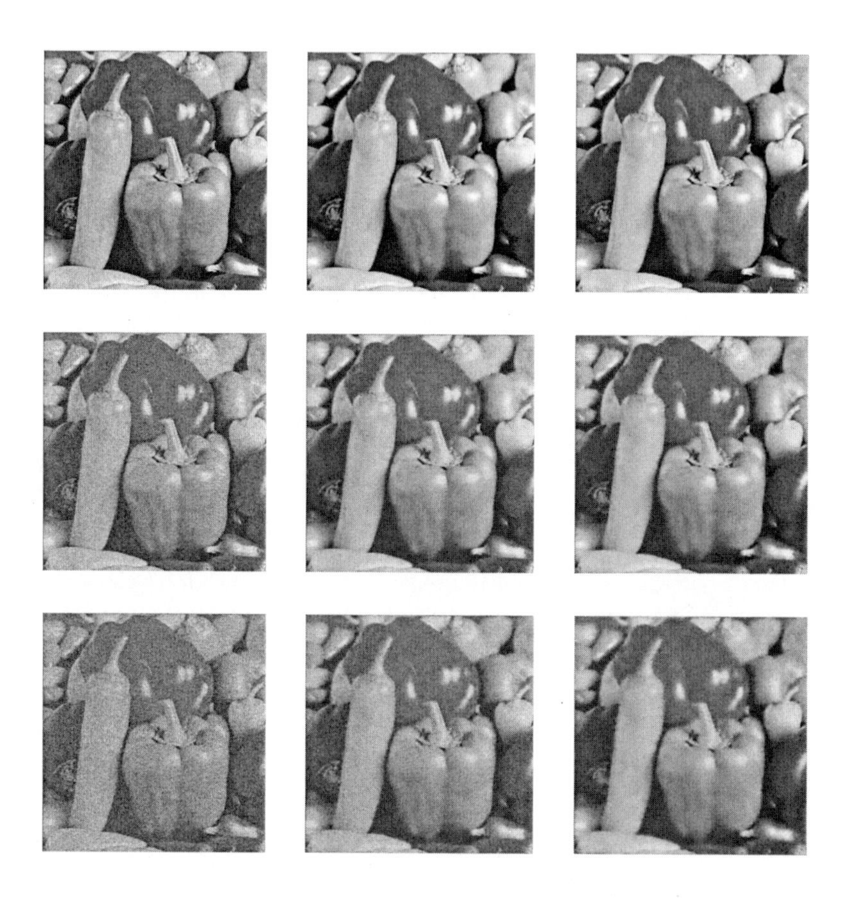

Figure 7.7: 'peppers', Gaussian white noise, from top to down: $\sigma_1 = 10, 25, 40$, from left to right: noisy, δ_H, δ_{MSE}

Figure 7.8: Salt&pepper noise, from left to right: noisy, δ_H, δ_{MSE}

Figure 7.9: Multiplicative Gaussian white noise, $\sigma_2 = 0.15$, from left to right: noisy, δ_H, δ_{MSE}

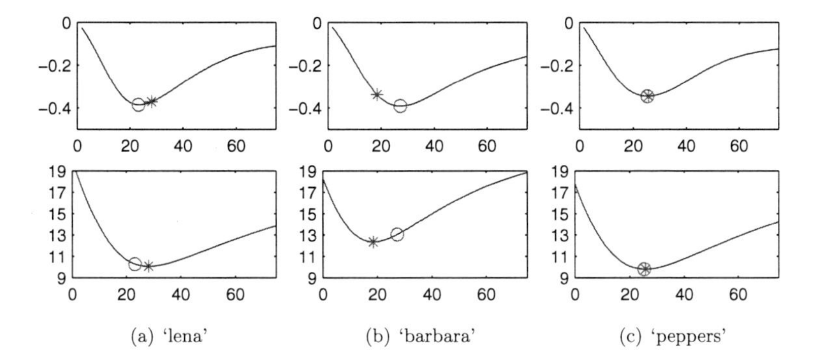

Figure 7.10: Multiplicative Gaussian white noise, $\sigma_2 = 0.15$, above: curvature/δ, below: RMSE/δ with δ_{MSE} ($*$) and δ_H ($\circ$)

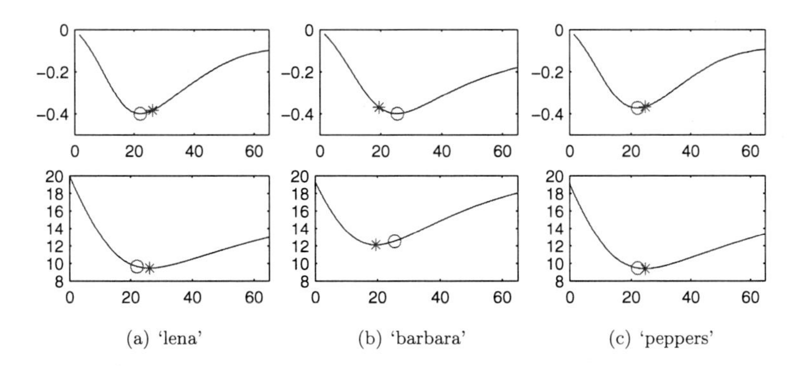

Figure 7.11: Additive and multiplicative Gaussian white noise, $\sigma_1 = 20$, $\sigma_2 = 0.1$, above: curvature/δ, below: RMSE/δ with δ_{MSE} ($*$) and δ_H ($\circ$)

Figure 7.12: Additive and multiplicative Gaussian white noise, $\sigma_1 = 20$, $\sigma_2 = 0.1$, from left to right: noisy, δ_H, δ_{MSE}

7.3.5 Summary of the Numerical Results

We corrupted three images 'lena', 'barbara', and 'peppers' by different kinds of noise. Then we compared the H-curve criterion for the bi-frame Laplace (2-2) with the MSE minimization. For medium and strong additive Gaussian white noise, the choice of the threshold parameter by the H-curve method provides better visual results than its choice with respect to a minimal MSE. Concerning salt&pepper noise, the H-curve criterion definitely outperforms the MSE minimization. For multiplicative as well as for a superposition of additive and multiplicative noise, both threshold choices yield satisfactory denoised images. In conclusion, we verified that the H-curve method is not only applicable to orthonormal wavelet bases but to bi-frames as well, and it provides very good results with respect to the visual perception. Finally, the present chapter indicates that bi-frames constitute a promising tool for applicational purposes.

Conclusion

In the present thesis, we studied multivariate wavelet frames with compact support. For a summary of our results, we revisit the problems (P1), (P2), (P3), and (P4) formulated in the introduction.

We have constructed smooth multivariate wavelet bi-frames with small supports and few wavelets, which are symmetric and have a high number of vanishing moments. Since primal and dual wavelets are derived from one single refinable function, they share exactly the same smoothness. The underlying refinable function is even fundamental, which is impossible within the concept of biorthogonal wavelets. Among others, we constructed two families of arbitrarily smooth bivariate wavelet bi-frames, Laplace (N_1-N_2) and Laplace (N_1-N_2)$_{\mathrm{R}}$, for the quincunx dilation matrix with three and two wavelets, respectively. Moreover, we derived two families of arbitrarily smooth wavelet bi-frames in arbitrary dimensions, Checkerboard (N) and Checkerboard (N)$_{\mathrm{R}}$, from three and two wavelets, respectively. The constructions we presented provide significantly smaller supports than comparable biorthogonal approaches. In conclusion, we have overcome the restrictions of biorthogonal wavelet bases, and we solved (P1) completely.

In order to verify that our wavelet bi-frames provide superior properties with respect to other wavelet frame constructions in the literature, we established a variety of optimality constraints addressing smoothness, approximation order, the number of wavelets, and symmetry. It turned out that the family Checkerboard (N)$_{\mathrm{R}}$ satisfies all of the optimality conditions. Hence, we also solved (P2) completely.

Next, we verified that Besov spaces are also characterized by wavelet bi-frames with isotropic scalings. We then addressed best N-term approximation, and by establishing matching Jackson and Bernstein inequalities and applying an interpolation result, we derived that the approximation classes of wavelet bi-frames with idempotent scalings are actually Besov spaces. Moreover, the best N-term approximation rate can be realized by an arbitrary thresholding rule applied to the bi-frame expansion. The findings provide a considerable extension of dyadic wavelet frame results in the literature, and they yield the solution to (P3). We also verified that the approximation classes of our multivariate wavelet bi-frames Checkerboard (N), (N)$_{\mathrm{R}}$ as well as the bivariate bi-frames Box Spline (1), (1)$_{\mathrm{R}}$, and (2)$_{\mathrm{R}}$ are Besov spaces. Unfortunately, the bi-frames Laplace (N_1-N_2) and Laplace (N_1-N_2)$_{\mathrm{R}}$ do not satisfy the assumptions for the Bernstein inequality. Hence, we cannot completely determine their approximation classes. Nevertheless, the Jackson inequality holds and thresholding the bi-frame expansion provides the predicted approximation rate.

Finally, we considered a variational problem for image denoising, and we discretized the problem with respect to the bi-frame Laplace (2-2). Then, we experimentally demonstrated that the H-curve method is also applicable to the wavelet bi-frame discretization and that it yields better results than the mean square error minimization with respect to the visual perception. Since the outcomes are promising, we solved (P4).

In the following, let us address some suggestions for future research. High-dimensional signal processing is a growing field, especially in medical imaging. With higher dimensions, the dyadic wavelet transform has a high complexity since the number of wavelets increases exponentially. One can circumvent these problems with dilation matrices satisfying $m = 2$, such as in (2.28). They also provide a much finer scaling, which may yield a better visual impression when changing the resolution by zooming in and out. The associated bi-frames Checkerboard (N), $(N)_{\mathrm{R}}$ are nonseparable and promising candidates for high-dimensional applications. Finally, the implementation of efficient high-dimensional algorithms for nonseparable wavelet bi-frames could be addressed in a future research project.

On the one hand, our constructive approaches in Chapters 3 and 4 lead to optimal wavelet bi-frames. On the other hand, the requirements of specific applications may often differ from the somehow abstract optimality criteria considered in the present thesis. In order to construct the most appropriate bi-frames, one must identify key properties for specific applications. This requires extensive numerical experiments, which might offer a second subject for future research.

In recent years, anisotropic Besov spaces have been the focus in the field of image and signal processing. In comparison to the Besov spaces we considered, they measure smoothness weighted in predefinable directions. Hence, they can express directional dependencies in an image, which is, for instance, advantageous in edge detection. In a third research project, one could attempt to characterize these spaces by wavelet bi-frames with general anisotropic dilation matrices. However, as far as we know, the characterization has not yet been completely clarified in terms of wavelet bases, see [GHT04, Hoc02] for some results on diagonal anisotropic matrices.

A Appendix

A.1 Function Spaces, Distributions, and the Fourier Transform

The Schwartz Space

The *Schwartz space* $\mathcal{S}(\mathbb{R}^d)$ is the collection of all $f \in \mathcal{C}^\infty(\mathbb{R}^d)$ such that

$$p_{\alpha,\beta}(f) := \sup_{x\in\mathbb{R}^d} \left| x^\alpha \partial^\beta f(x) \right|$$

is finite for all $\alpha, \beta \in \mathbb{N}_0^d$. Equipped with the collection of semi-norms $\{p_{\alpha,\beta} : \alpha, \beta \in \mathbb{N}_0^d\}$, it constitutes a Fréchet space, cf. [Trè67] for details.

The Fourier Transform and Tempered Distributions

For $f \in L_1(\mathbb{R}^d) \cap L_2(\mathbb{R}^d)$, let

$$\widehat{f}(\xi) := \int_{\mathbb{R}^d} f(x) e^{-2\pi i x\cdot\xi} dx,$$

denote the *Fourier transform* of f. We also apply the notation $\mathcal{F}f := \widehat{f}$. Then $\mathcal{F}$ is an automorphism of $\mathcal{S}(\mathbb{R}^d)$, and it extends to an isometric automorphism on $L_2(\mathbb{R}^d)$.

The Schwartz space's topological dual space $\mathcal{S}'(\mathbb{R}^d)$ is called the collection of *tempered distributions*. Then $\mathcal{F}$ extends to an automorphism of $\mathcal{S}'(\mathbb{R}^d)$ by

$$(\mathcal{F}f)(\eta) := f(\mathcal{F}\eta), \quad \text{for } f \in \mathcal{S}'(\mathbb{R}^d) \text{ and } \eta \in \mathcal{S}(\mathbb{R}^d).$$

For more details about distributions and the Fourier transform as well as for the support of a distribution, their translation, dilation, and convolution, we refer to [Trè67].

Sobolev Spaces

Let Ω be an arbitrary domain in $\mathbb{R}^d$. For $s \in \mathbb{N}_0$ and $1 \le p \le \infty$, denote

$$W^s(L_p(\Omega)) := \{f \in L_p(\Omega) : \partial^\alpha f \in L_p(\Omega), \text{ for all } |\alpha| \le s\}$$

the *Sobolev space* of smoothness s in $L_p(\Omega)$, where $\partial^\alpha f$ means α weak partial derivatives of f. Equipped with the norm

$$\|f\|_{W_p^s(\Omega)} := \sum_{|\alpha|\le s} \|\partial^\alpha f\|_{L_p}$$

it is a Banach space. As required in Theorem 5.3.4, we denotes by

$$|f|_{W_p^s(\Omega)} := \sum_{|\alpha|= s} \|\partial^\alpha f\|_{L_p}$$

the *Sobolev semi-norm*.

Next, we address fractional s. For $0 < s \notin \mathbb{N}$, let $s = k + \nu$ with $k \in \mathbb{N}_0$ and $0 < \nu < 1$. Then for $1 \leq p < \infty$, the *fractional Sobolev space* is given by

$$W^s(L_p(\Omega)) := \left\{ f \in W^k(L_p(\Omega)) : \|f\|_{W^s(L_p)} < \infty \right\},$$

where

$$\|f\|^p_{W^s(L_p)} := \|f\|^p_{W^k(L_p)} + \sum_{|\alpha|=k} \int_\Omega \int_\Omega \frac{|\partial^\alpha f(x) - \partial^\alpha f(y)|^p}{|x-y|^{d+\nu p}} dx dy.$$

Hölder Spaces

Denote $\mathcal{C}(\mathbb{R}^d)$ the collection of all bounded and uniformly continuous functions $f : \mathbb{R}^d \to \mathbb{C}$. For $s \in \mathbb{N}_0$, let

$$\mathcal{C}^s(\mathbb{R}^d) := \left\{ f : \partial^\alpha f \in \mathcal{C}(\mathbb{R}^d), \text{ for all } |\alpha| \leq s \right\}.$$

The smoothness parameter s can be extended to fractional $s > 0$: let k be an integer such that $s = k + \nu$, $0 \leq \nu < 1$, then the *Hölder space* $\mathcal{C}^s(\mathbb{R}^d)$ is the collection of all $f \in \mathcal{C}^k(\mathbb{R}^d)$ such that

$$|\partial^\alpha f(x+h) - \partial^\alpha f(x)| \lesssim \|h\|^\nu, \text{ for all } h \in \mathbb{R}^d,\ |\alpha| = k.$$

Lipschitz Spaces

Given $f : \mathbb{R}^d \to \mathbb{C}$ and $h \in \mathbb{R}^d$, denote

$$\Delta^1_h f := f(\cdot + h) - f(\cdot) \tag{A.1}$$

the *difference of order* 1. For $f \in L_p(\mathbb{R}^d)$, $0 < p \leq \infty$, and $t \geq 0$, the *modulus of continuity* is defined by

$$\omega(f,t)_{L_p} := \sup_{|h| \leq t} \left\| \Delta^1_h f \right\|_{L_p}. \tag{A.2}$$

The *Lipschitz space* $\mathrm{Lip}\left(s, L_p(\mathbb{R}^d)\right)$, for $0 < s \leq 1 \leq p \leq \infty$, consists of all $f \in L_p(\mathbb{R}^d)$ such that

$$\omega(f,t)_{L_p} \lesssim t^s, \quad \text{for all } t > 0.$$

For $s > 1$, let k be an integer such that $s = k + \nu$, $0 < \nu \leq 1$. The Lipschitz space extends to the range $s > 1$ by

$$\mathrm{Lip}\left(s, L_p(\mathbb{R}^d)\right) := \left\{ f \in L_p(\mathbb{R}^d) : \partial^\alpha f \in \mathrm{Lip}\left(\nu, L_p(\mathbb{R}^d)\right), \text{ for all } |\alpha| = k \right\}.$$

For $f \in L_p(\mathbb{R}^d)$, we call

$$s_p(f) := \sup\left\{ s \geq 0 : f \in \mathrm{Lip}\left(s, L_p(\mathbb{R}^d)\right) \right\}$$

its *L_p-critical exponent.* Due to [DL93], provided that $s \in \mathbb{N}$ and $1 < p \leq \infty$, the Lipschitz space equals the Sobolev space, i.e.,

$$\mathrm{Lip}\left(s, L_p(\mathbb{R}^d)\right) = W^s(L_p(\mathbb{R}^d)). \tag{A.3}$$

Thus, $s_\infty(f)$ provides a tool for verifying membership in $W^s(L_\infty(\mathbb{R}^d))$: if $0 < s_\infty(f) = k + \nu$, k an integer and $0 < \nu \le 1$, then $f \in W^k(L_\infty(\mathbb{R}^d))$.

For all $0 < \varepsilon < s$, the fractional Sobolev spaces provide the following inclusions

$$W^s(L_2(\mathbb{R}^d)) \subset \mathrm{Lip}\left(s, L_2(\mathbb{R}^d)\right) \subset W^{s-\varepsilon}(L_2(\mathbb{R}^d)), \tag{A.4}$$

see for example [Jia99]. This yields

$$s_2(f) = \left\{s \ge 0 : f \in W^s(L_2(\mathbb{R}^d))\right\}.$$

Next, we derive an interpretation of $s_\infty(f)$: if $s > 0$ is not an integer, then

$$\mathrm{Lip}\left(s, L_\infty(\mathbb{R}^d)\right) = \mathcal{C}^s(\mathbb{R}^d),$$

see [DL93]. Thus, the L_∞-critical exponent $s_\infty(f)$ measures Hölder smoothness, i.e., for $f \in L_\infty(\mathbb{R}^d)$,

$$s_\infty(f) = \left\{s \ge 0 : f \in \mathcal{C}^s(\mathbb{R}^d)\right\}. \tag{A.5}$$

The Hardy Space

Following [FS72], we briefly recall Hardy spaces. Given $f \in \mathcal{S}'(\mathbb{R}^d)$ and $\phi \in \mathcal{S}(\mathbb{R}^d)$ with $\widehat{\phi}(0) = 1$, let

$$f^\star(x) := \sup_{|x-y|<t} \left|\mathcal{F}^{-1}\left(\phi(t\cdot)\mathcal{F}f\right)(y)\right|.$$

Then the *Hardy space* $H_p(\mathbb{R}^d)$, for $0 < p < \infty$, is the collection of all $f \in \mathcal{S}'(\mathbb{R}^d)$ such that $\|f^\star\|_{L_p}$ is finite. It turns out that $H_p(\mathbb{R}^d) = L_p(\mathbb{R}^d)$, for $1 < p < \infty$, see [FS72, Tri92] for this result and for a more detailed discussion about Hardy spaces.

Functions of Bounded Variation

We only require the space of bivariate functions of bounded variation, and we follow the approach in [CDPX99]. Given $f \in L_1(\mathbb{R}^2)$, let

$$\mathrm{v}(f) := \sup_{0<h} \frac{1}{h} \sum_{j=1}^{2} \left\|\Delta^1_{he_j} f\right\|_{L_1}$$

denote the *variation* of f, where e_1, e_2 denote the two coordinate vectors in $\mathbb{R}^2$ and Δ^1 is the difference of order one as in (A.1). Then $BV(\mathbb{R}^2)$ is the collection of all $f \in L_1(\mathbb{R}^2)$ such that the norm

$$\|f\|_{BV} := \mathrm{v}(f) + \|f\|_{L_1}$$

is finite. For details about $BV(\mathbb{R}^2)$, we refer to [CDPX99] and [Mey01].

A.2 Some (Alternative) Proofs

Proof of Lemma 1.1.3

Lemma. *Each idempotent dilation matrix is isotropic.*

Proof. Let $l, h \in \mathbb{N}$ with $M^l = h\mathcal{I}_d$. Since each eigenvalue ρ_i of M satisfies $\rho_i^l = h$, they have the same modulus, i.e., $|\rho_i| = \rho$, $i = 1, \dots, d$. Next, we address the diagonalization. Since $M^l - h\mathcal{I}_d = 0$, the minimum polynomial of M is a divisor of the polynomial $x^l - h$. Hence, it has pairwise distinct zeros, which provides that M can be diagonalized, see a standard textbook on linear algebra. □

Proof of Lemma 1.3.1

Lemma. *Given a pair of compactly supported biorthogonal wavelet bases, let their associated refinable functions φ and $\widetilde{\varphi}$ be generated by symbols a and b, respectively, with $a(0) = b(0) = 1$. If both refinable functions are fundamental, then $m > 2$.*

Proof. We suppose $m = 2$. The biorthogonality of φ and $\widetilde{\varphi}$ implies that b is a dual symbol of a. Hence, their subsymbols satisfy

$$A_0\overline{B_0} + A_{\gamma_1^*}\overline{B_{\gamma_1^*}} = 2,$$

see Subsection 1.1.4. Since both φ and $\widetilde{\varphi}$ are fundamental, their subsymbols A_0 and B_0 are equal to 1, cf. [DM97]. This yields

$$A_{\gamma_1^*}\overline{B_{\gamma_1^*}} = 1,$$

which implies

$$A_{\gamma_1^*}(\xi) = B_{\gamma_1^*}(\xi) = te^{-2\pi i l\cdot\xi},$$

for some $l \in \mathbb{Z}^d$ and $t \in \mathbb{C}$ with $|t| = 1$. The normalization $a(0) = b(0) = 1$ yields $t = 1$. So far, we have

$$a(\xi) = b(\xi) = \frac{1}{2}\left(1 + e^{-2\pi i(Ml+\gamma_1^*)\cdot\xi}\right).$$

According to [GM92], the generated refinable functions of such symbols are characteristic functions. Hence, φ and $\widetilde{\varphi}$ are not continuous, which contradicts that they are fundamental. □

Proof of Corollary 1.2.5

Corollary. *Given an idempotent dilation matrix M, let φ be a compactly supported continuous refinable function in $L_2(\mathbb{R}^d)$ with $\widehat{\varphi}(0) \neq 0$. Then φ reproduces polynomials up to order s iff its multiresolution analysis provides approximation order s.*

Proof. Since idempotent dilation matrices are isotropic, one direction of the equivalence is provided by Theorem 1.2.4. In order to address the reverse implication, let the underlying multiresolution analysis provide approximation order s, i.e.,

$$\operatorname{dist}(f, V_j)_{L_2} \lesssim \rho^{-js}. \tag{A.6}$$

Let $l, h \in \mathbb{N}$ such that $M^l = h\mathcal{I}_d$. Applying l-times the refinement equation yields that φ is also refinable with respect to $h\mathcal{I}_d$. Hence, φ also generates the multiresolution analysis $(V_{lj})_{j\in\mathbb{Z}}$ with scaling $h\mathcal{I}_d$. Obviously, (A.6) and $\rho^l = h$ imply

$$\operatorname{dist}(f, V_{lj})_{L_2} \lesssim \rho^{-ljs} = h^{-js}.$$

Hence, $(V_{lj})_{j\in\mathbb{Z}}$ provides approximation order s. By applying Theorem 1.2.3, we conclude the proof. □

Proof of Lemma 2.1.6

Lemma. *Given a frame* $\{f_\kappa : \kappa \in \mathcal{K}\}$ *for* $\mathcal{H}$*, let* $\{\widetilde{f}_\kappa : \kappa \in \mathcal{K}\} \subset \mathcal{H}$ *be a biorthogonal sequence, i.e.,*

$$\left\langle f_\kappa, \widetilde{f}_{\kappa'} \right\rangle = \delta_{\kappa,\kappa'}, \quad \text{for all } \kappa, \kappa' \in \mathcal{K}.$$

Then both systems constitute a pair of biorthogonal Riesz bases for $\mathcal{H}$.

Proof. The synthesis operator

$$F : \ell_2(\mathcal{K}) \to \mathcal{H}$$

of the frame $\{f_\kappa : \kappa \in \mathcal{K}\}$ is well-defined and onto. Due to the biorthogonality, it is also injective. Hence, $\{f_\kappa : \kappa \in \mathcal{K}\}$ constitutes a Riesz basis. In the sequel, we verify that the collection $\{\widetilde{f}_\kappa : \kappa \in \mathcal{K}\}$ is also a Riesz basis. Given $f \in \mathcal{H}$, we have

$$f = \sum_{\kappa \in \mathcal{K}} c_\kappa f_\kappa,$$

for a sequence $(c_\kappa)_{\kappa\in\mathcal{K}} \in \ell_2(\mathcal{K})$. According to the biorthogonality relations, the sequence of inner products $\left(\langle f, \widetilde{f}_\kappa\rangle\right)_{\kappa\in\mathcal{K}}$ equals $(c_\kappa)_{\kappa\in\mathcal{K}}$, which means it is also contained in $\ell_2(\mathcal{K})$. Then the analysis operator

$$\widetilde{F}^* : \mathcal{H} \to \ell_2(\mathcal{K}), \quad f \mapsto (\langle f, \widetilde{f}_\kappa\rangle)_{\kappa\in\mathcal{K}}$$

is well-defined, and so is the synthesis operator

$$\widetilde{F} : \ell_2(\mathcal{K}) \to \mathcal{H}, \quad (c_\kappa)_{\kappa\in\mathcal{K}} \mapsto \sum_{\kappa\in\mathcal{K}} c_\kappa \widetilde{f}_\kappa.$$

Thus, $\{\widetilde{f}_\kappa : \kappa \in \mathcal{K}\}$ is at least a Bessel sequence. By applying the biorthogonality, we have

$$\left\langle f - \sum_{\kappa\in\mathcal{K}} \langle f, f_\kappa\rangle \widetilde{f}_\kappa \, , \, f_{\kappa'} \right\rangle = 0, \quad \text{for all } \kappa' \in \mathcal{K}.$$

Since $\{f_\kappa : \kappa \in \mathcal{K}\}$ is complete in $\mathcal{H}$, the synthesis operator $\widetilde{F}$ is onto, which means that the collection $\{\widetilde{f}_\kappa : \kappa \in \mathcal{K}\}$ is a frame. Then by repeating the above arguments, we can conclude the proof. □

An Alternative Proof of Corollary 4.2.6

Corollary (Step 1). *Using the notation of Theorem* 4.1.3*, let the symbols* $a^{(0)}$ *and* $b^{(0)}$ *satisfy the sum rules of order* $2s$ *and* $a^{(0)}(0) = b^{(0)}(0) = 1$. *Additionally, let* $a^{(0)}$ *be interpolatory. For* $\mu = 1, \ldots, m-1$*, we define*

$$a^{(\mu)}(\xi) = e^{-2\pi i \gamma_\mu^* \cdot \xi} \sum_{\gamma\in\Gamma_M\setminus\{0\}} \left(a^{(0)}(\xi+\gamma) - a^{(0)}(\xi) e^{-2\pi i\gamma_\mu^*\cdot\gamma}\right) \overline{b^{(0)}(\xi+\gamma)}, \tag{A.7}$$

$$b^{(\mu)}(\xi) = \frac{1}{m} e^{-2\pi i \gamma_\mu^* \cdot \xi} \sum_{\gamma\in\Gamma_M\setminus\{0\}} \left(1 - e^{-2\pi i\gamma_\mu^*\cdot\gamma}\right) \overline{a^{(0)}(\xi+\gamma)}. \tag{A.8}$$

Then (S1) *in Theorem* 4.1.3 *is satisfied, and* (S1*) *holds for* $s_1 = 2s$.

Proof. Given $a^{(\mu)}$ and $b^{(\mu)}$, $\mu = 1,\dots,m-1$, as in (A.7) and (A.8), respectively. A direct calculation yields

$$a^{(\mu)}(\xi) = e^{-2\pi i\gamma_\mu^*\cdot\xi}\Big(\overline{\theta(\xi)} - a^{(0)}(\xi)\sum_{\gamma\in\Gamma_M} e^{-2\pi i\gamma_\mu^*\cdot\gamma}\overline{b^{(0)}(\xi+\gamma)}\Big). \tag{A.9}$$

Since $a^{(0)}$ is interpolatory, we obtain

$$b^{(\mu)}(\xi) = \frac{1}{m}e^{-2\pi i\gamma_\mu^*\cdot\xi}\Big(1 - \sum_{\gamma\in\Gamma_M} e^{-2\pi i\gamma_\mu^*\cdot\gamma}\overline{a^{(0)}(\xi+\gamma)}\Big). \tag{A.10}$$

The following change of notation simplifies the remainder of the proof. Let us denote $a := a^{(0)}$ and $b := b^{(0)}$. Let then $a^{(0)}$ and $b^{(0)}$ be redefined by (A.9), (A.10) for $\mu = 0$, while a and b remain unchanged. Note that then the interpolation condition yields $b^{(0)} = 0$, which trivially implies

$$\sum_{\mu=1}^{m-1}\overline{a^{(\mu)}(\xi+\gamma)}b^{(\mu)}(\xi) = \sum_{\mu=0}^{m-1}\overline{a^{(\mu)}(\xi+\gamma)}b^{(\mu)}(\xi). \tag{A.11}$$

Applying the identity (1.41) yields

$$\begin{aligned}
\sum_{\mu=0}^{m-1}\overline{a^{(\mu)}(\xi+\gamma)}b^{(\mu)}(\xi) &= \frac{1}{m}\sum_{\mu=0}^{m-1}\Big(e^{2\pi i\gamma_\mu^*\cdot\gamma}\theta(\xi) - \theta(\xi)\sum_{\tilde\gamma\in\Gamma_M} e^{-2\pi i\gamma_\mu^*\cdot(\tilde\gamma-\gamma)}\overline{a(\xi+\tilde\gamma)}\\
&\quad - \overline{a(\xi+\gamma)}\sum_{\tilde\gamma\in\Gamma_M} e^{2\pi i\gamma_\mu^*\cdot(\tilde\gamma+\gamma)}b(\xi+\gamma+\tilde\gamma)\\
&\quad + \overline{a(\xi+\gamma)}\sum_{\tilde\gamma,\tilde{\tilde\gamma}\in\Gamma_M} e^{2\pi i\gamma_\mu^*\cdot(\tilde\gamma+\gamma-\tilde{\tilde\gamma})}b(\xi+\gamma+\tilde\gamma)\overline{a(\xi+\tilde{\tilde\gamma})}\Big)\\
&= \delta_{0,\gamma}\theta(\xi) - \theta(\xi)\overline{a(\xi+\gamma)} - \overline{a(\xi+\gamma)}b(\xi)\\
&\quad + \overline{a(\xi+\gamma)}\sum_{\tilde{\tilde\gamma}\in\Gamma_M} b(\xi+\tilde{\tilde\gamma})\overline{a(\xi+\tilde{\tilde\gamma})}\\
&= \delta_{0,\gamma}\theta(\xi) - \theta(\xi)\overline{a(\xi+\gamma)} - \overline{a(\xi+\gamma)}b(\xi) + \overline{a(\xi+\gamma)}\theta(\xi)\\
&= \delta_{0,\gamma}\theta(\xi) - \overline{a(\xi+\gamma)}b(\xi).
\end{aligned}$$

□

Since $b^{(0)} = 0$, this concludes the proof.

Proof of Theorem 5.2.5

Theorem. *Under the assumptions of Proposition 5.2.3, let $\varphi \in W^s(L_\infty(\mathbb{R}^d))$, $s \in \mathbb{N}$, and suppose $\widehat{\varphi}(0) \neq 0$. Then, for $0 < \alpha < s$ and for all $f \in L_p(\mathbb{R}^d)$,*

$$\|f\|_{\dot B_q^\alpha(L_p)} \sim \Big\|\big(\rho^{\alpha j}\operatorname{dist}(f,V_j)_{L_p}\big)_{j\in\mathbb{Z}}\Big\|_{\ell_q}. \tag{A.12}$$

For preparation, we need the following Hardy-type inequalities:

Lemma A.2.1. *Let* $(x_j)_{j\in\mathbb{Z}}$, $(y_j)_{j\in\mathbb{Z}}$ *be nonnegative real numbers and let* $u>1$ *be fixed. Given* $0<\beta<\infty$ *such that*

$$x_j \lesssim u^{-j\beta} \sum_{i=-\infty}^{j} u^{i\beta} y_i, \quad \text{for all } j \in \mathbb{Z},$$

then, for all $0<\alpha<\beta$ *and* $1\le\tau\le\infty$,

$$\left\| \left(u^{j\alpha} x_j\right)_{j\in\mathbb{Z}} \right\|_{\ell_\tau} \lesssim \left\| \left(u^{j\alpha} y_j\right)_{j\in\mathbb{Z}} \right\|_{\ell_\tau}.$$

Proof. For $u=2$, a proof is contained in [Jia93]. With the result for $u=2$ in hand, the general statement follows by the changeover from β and α to $\beta\log_2(u)$ and $\alpha\log_2(u)$, respectively. □

In the following proof, we combine Proposition 5.2.3 and Lemma A.2.1. It extends the dyadic results in [DJP92]:

Proof of the Theorem. Since φ is compactly supported, it is also contained in the space $W^s(L_1(\mathbb{R}^d))$. Then according to Theorem 1.2.6, it reproduces polynomials up to order $s+1$, and one direction of the equivalence follows directly by applying (5.22). Next, we address the reverse estimate. Given $f\in L_p(\mathbb{R}^d)$, let $f_j\in V_j$ such that

$$\|f-f_j\|_{L_p} \le 2\,\mathrm{dist}(f,V_j)_{L_p}. \tag{A.13}$$

For the range $0<p<1$, the space $L_p(\mathbb{R}^d)$ is only quasi-normed. Nevertheless, we have

$$\|f+g\|_{L_p}^{\vartheta} \le \|f\|_{L_p}^{\vartheta} + \|g\|_{L_p}^{\vartheta}, \tag{A.14}$$

for the complete range $0<p<\infty$, where $\vartheta=\min(1,p)$, cf. Chapter 2 in [DL93]. Given $0<\vartheta\le 1$, the mapping $t\mapsto t^{\vartheta}$ is concave on the positive real line. Hence, the inequality (A.14) holds for all $0<\vartheta\le\min(1,p)$, which provides

$$\omega_l\left(f+g,t\right)_{L_p}^{\vartheta} \le \omega_l\left(f,t\right)_{L_p}^{\vartheta} + \omega_l\left(g,t\right)_{L_p}^{\vartheta}. \tag{A.15}$$

Let us fix ϑ by

$$\vartheta := \min\{1,p,q\}.$$

Note that (A.13), (A.14), and $0\in V_j$ imply

$$\begin{aligned}\|f_j\|_{L_p}^{\vartheta} &\le \|f_j-f\|_{L_p}^{\vartheta} + \|f\|_{L_p}^{\vartheta}\\ &\le 2^{\vartheta}\,\mathrm{dist}(f,V_j)_{L_p}^{\vartheta} + \|f\|_{L_p}^{\vartheta}\\ &\le (2^{\vartheta}+1)\|f\|_{L_p}^{\vartheta},\end{aligned}$$

which yields

$$\|f_j\|_{L_p}^{\vartheta} \lesssim \|f\|_{L_p}^{\vartheta}. \tag{A.16}$$

For $j_0<j$, the application of (A.15) provides

$$\omega_l(f,\rho^{-j})_{L_p}^{\vartheta} \le \omega_l(f-f_j,\rho^{-j})_{L_p}^{\vartheta} + \sum_{i=j_0}^{j} \omega_l(f_i-f_{i-1},\rho^{-j})_{L_p}^{\vartheta} + \omega_l(f_{j_0-1},\rho^{-j})_{L_p}^{\vartheta}. \tag{A.17}$$

Since, for $s < l \in \mathbb{N}$, the inequality (5.23) implies

$$\omega_l(f,\rho^{-j})_{L_p} \lesssim \min\{1,\rho^{j_0-j}\}^s \|f\|_{L_p}, \quad \text{for all } f \in V_{j_0},\ j,j_0 \in \mathbb{Z}, \tag{A.18}$$

we obtain with (A.16)

$$\begin{aligned}\omega_l(f_{j_0-1},\rho^{-j})_{L_p}^{\vartheta} &\lesssim \rho^{s(j_0-1-j)\vartheta}\, \|f_{j_0-1}\|_{L_p}^{\vartheta} \\ &\lesssim \rho^{s(j_0-1-j)\vartheta}\, \|f\|_{L_p}^{\vartheta},\end{aligned}$$

which converges to zero as j_0 goes to minus infinity. With the rough estimate

$$\omega_l(f,t)_{L_p} \lesssim \|f\|_{L_p},$$

we obtain from (A.17)

$$\omega_l(f,\rho^{-j})_{L_p}^{\vartheta} \lesssim \|f-f_j\|_{L_p}^{\vartheta} + \sum_{i=-\infty}^{j} \omega_l(f_i - f_{i-1},\rho^{-j})_{L_p}^{\vartheta}.$$

By applying $f_i - f_{i-1} \in V_i$ and (A.18), this yields

$$\begin{aligned}\omega_l(f,\rho^{-j})_{L_p}^{\vartheta} &\lesssim \|f-f_j\|_{L_p}^{\vartheta} + \sum_{i=-\infty}^{j} \rho^{(i-j)s\vartheta}\|f_i - f_{i-1}\|_{L_p}^{\vartheta} \\ &\leq \|f-f_j\|_{L_p}^{\vartheta} + \sum_{i=-\infty}^{j} \rho^{(i-j)s\vartheta}\|f_i - f\|_{L_p}^{\vartheta} + \sum_{i=-\infty}^{j} \rho^{(i-j)s\vartheta}\|f - f_{i-1}\|_{L_p}^{\vartheta} \\ &= \|f-f_j\|_{L_p}^{\vartheta} + \sum_{i=-\infty}^{j} \rho^{(i-j)s\vartheta}\|f_i - f\|_{L_p}^{\vartheta} + \sum_{i=-\infty}^{j-1} \rho^{(i+1-j)s\vartheta}\|f - f_i\|_{L_p}^{\vartheta} \\ &\lesssim \sum_{i=-\infty}^{j} \rho^{(i-j)s\vartheta}\|f - f_i\|_{L_p}^{\vartheta},\end{aligned}$$

where a factor $\rho^{s\vartheta}$ is contained in the constant, in order to derive the last estimate. According to (A.13), we have

$$\omega_l(f,\rho^{-j})_{L_p}^{\vartheta} \lesssim \rho^{-js\vartheta} \sum_{i=-\infty}^{j} \rho^{is\vartheta}\operatorname{dist}(f,V_i)_{L_p}^{\vartheta}.$$

Now, we apply the Hardy inequalities of Lemma A.2.1 with

$$\begin{aligned}x_j &:= \omega_l(f,\rho^{-j})_{L_p}^{\vartheta}, \\ y_i &:= \operatorname{dist}(f,V_i)_{L_p}^{\vartheta},\end{aligned}$$

as well as $u=\rho$, $\beta = s\vartheta$, and $\tau = \frac{q}{\vartheta} \geq 1$. This concludes the proof. □

Proof of Lemma 5.3.9

Lemma. *Let a finite number of compactly supported functions $\psi^{(\mu)}$, $\mu = 1,\dots,n$, be given. Then their dilates and shifts satisfy the following overlapping condition:*

(a) $\left|\operatorname{supp}\left(\psi_{j,k}^{(\mu)}\right)\right| \lesssim m^{-j}$.

(b) *Let μ and k be fixed. Then, for all $j \in \mathbb{Z}^d$,*

$$\operatorname{card}\left\{(\nu, l) : \operatorname{supp}\left(\psi_{j,k}^{(\mu)}\right) \cap \operatorname{supp}\left(\psi_{j,l}^{(\nu)}\right) \neq \emptyset\right\} \lesssim 1. \tag{A.19}$$

Proof. (a): One easily verifies

$$\operatorname{supp}\left(\psi_{j,k}^{(\mu)}\right) = M^{-j}\left(\operatorname{supp}\left(\psi^{(\mu)}\right) + k\right).$$

Then for

$$C := \max_{\mu=1,\dots,n}\left|\operatorname{supp}\psi^{(\mu)}\right|,$$

we obtain

$$\left|\operatorname{supp}\left(\psi_{j,k}^{(\mu)}\right)\right| = m^{-j}\left|\operatorname{supp}\left(\psi^{(\mu)}\right)\right| \le Cm^{-j}.$$

(b): On the scale $j = 0$, there is only a finit number of overlappings because of the compact support. On scale j, we have

$$\begin{aligned}
\operatorname{supp}\left(\psi_{j,k}^{(\mu)}\right) \cap \operatorname{supp}\left(\psi_{j,l}^{(\nu)}\right) &= M^{-j}\left(\operatorname{supp}\left(\psi^{(\mu)}\right) + k\right) \cap M^{-j}\left(\operatorname{supp}\left(\psi^{(\nu)}\right) + l\right) \\
&= M^{-j}\left(\left(\operatorname{supp}\left(\psi^{(\mu)}\right) + k\right) \cap \left(\operatorname{supp}\left(\psi^{(\nu)}\right) + l\right)\right),
\end{aligned}$$

where the last equality is valid because M^{-j} is an injective mapping. Thus, the number of overlappings does not depend on the scale. □

Proof of Lemma 5.3.14

Lemma. *Let $1 \le q < p \le \infty$ and let $f_n \in L_p(\mathbb{R}^d) \cap L_q(\mathbb{R}^d)$ converge to f in $L_p(\mathbb{R}^d)$ and to g in $L_q(\mathbb{R}^d)$. Then $f = g$ up to a set of measure zero.*

Proof. Let $K \subset \mathbb{R}^d$ be compact. Then $f_n \cdot 1_K$ converges to $f \cdot 1_K$ in $L_p(\mathbb{R}^d)$ and to $g \cdot 1_K$ in $L_q(\mathbb{R}^d)$, where 1_K denotes the characteristic function of the set K. By applying the embedding

$$L_p(K) \hookrightarrow L_q(K),$$

we obtain

$$\begin{aligned}
\|f - g\|_{L_q(K)} &\le \|f - f_n\|_{L_q(K)} + \|f_n - g\|_{L_q(K)} \\
&\lesssim \|f - f_n\|_{L_p(K)} + \|f_n - g\|_{L_q(K)}.
\end{aligned}$$

The right-hand side goes to zero as n tends to infinity. Thus, $f = g$ on K up to a set of measure zero. Finally, the space $\mathbb{R}^d$ can be exhausted by a countable collection of compact sets. This concludes the proof. □

An Alternative Proof of Proposition 6.4.11

In the sequel, we present a proof of Proposition 6.4.11 in terms of masks. Similar to linear independence on $(0,1)^d$, local linear independence can also be expressed in terms of the underlying mask, cf. [Ron99]:

Theorem A.2.2. *Let a symbol a with $\operatorname{supp}(a_k)_{k\in\mathbb{Z}^d} \subset [0,L]^d$ generate a continuous refinable functions φ with respect to dyadic dilation. Then φ has locally linearly independent integer shifts iff, for all $0 \le \nu \le 2^{N^d}$ and $\gamma_1^*, \ldots, \gamma_\nu^* \in \{0,1\}^d$, the nonzero rows of the matrix $\mathscr{A}_{\gamma_1^*} \cdots \mathscr{A}_{\gamma_\nu^*}\mathcal{M}$ are linearly independent.*

It should be mentioned that Theorem A.2.2 seems useless for practical computations. Especially in the multivariate setting, the dimensions of the matrices become very large, and since ν varies between 0 and $2^{(L^d)}$, the number of matrices to be checked explodes. We avoid such difficulties since we merely apply the theorem in a theoretical sense. Before we can do so, we still need preparation. Let $\odot$ denote the *Kronecker product* of matrices, i.e., for $A = (\alpha_{i,j})$ and B some matrices of arbitrary dimensions with entries in $\mathbb{C}$,

$$A \odot B := (\alpha_{i,j} B)\,.$$

According to [Duf56], we have the two relations

$$\operatorname{rank}(A \odot B) = \operatorname{rank}(A) \cdot \operatorname{rank}(B), \tag{A.20}$$

$$(A \odot B)\,(C \odot D) = (AC) \odot (BD)\,, \tag{A.21}$$

where C, D should be matrices with compatible dimensions. The application of (A.20) and (A.21) yield the following lemma:

Lemma A.2.3. *Given some univariate dyadic symbol a with $(a_k)_{k\in\mathbb{Z}} \subset [0,L]$, let it generate a continuous refinable function φ. Then let $\widetilde{a} = \bigotimes_{i=1}^d a$ be the d-dimensional tensor symbol considered with respect to dyadic dilation. Moreover, let $\widetilde{K} := [0,L]^d$ be ordered by $\{(0,\ldots,0)^\top, (0,\ldots,0,1)^\top, \ldots\}$. Then the following holds:*

(a) *Let $\gamma^* = (\gamma_1^*, \ldots, \gamma_d^*)^\top \in \{0,1\}^d$ then $\widetilde{\mathscr{A}_{\gamma^*}} = \bigodot_{i=1}^d \mathscr{A}_{\gamma_i^*}$.*

(b) *The vector $\widetilde{\Phi}$ is given by $\widetilde{\Phi}(x) = \bigodot_{i=1}^d \Phi(x_i)$.*

(c) *The matrix $\widetilde{\mathcal{M}}$ can be chosen by $\widetilde{\mathcal{M}} = \bigodot_{i=1}^d \mathcal{M}$.*

Proof. (a) One simply verifies the following equalities:

$$\begin{aligned}\widetilde{\mathscr{A}_{\gamma^*}} &= \left(\widetilde{a}_{\gamma^*+2k-k'}\right)_{k,k'\in\widetilde{K}} = \left(\prod_{i=1}^d a_{\gamma_i^*+2k_i-k_i'}\right)_{k,k'\in\widetilde{K}} \\ &= \bigodot_{i=1}^d \left(a_{\gamma_i^*+2k_i-k_i'}\right)_{k,k'\in K} = \bigodot_{i=1}^d \mathscr{A}_{\gamma_i^*}\,.\end{aligned}$$

(b) The refinable function of $\widetilde{a}$ is given by the tensor product $\bigotimes_{i=1}^d \varphi$, cf. [Dau92]. Then we have

$$\widetilde{\Phi}(x) = (\widetilde{\varphi}(x-k))_{k\in\widetilde{K}} = \left(\bigotimes_{i=1}^d \varphi(x_i - k_i)\right)_{k\in\widetilde{K}}\,.$$

Finally, the last term equals $\bigodot_{i=1}^d \Phi(x_i)$.

(c) The columns of $\mathcal{M}$ constitute a basis for the minimal common invariant subspace $\mathcal{V}$ of $\{\mathscr{A}_{\gamma^*} : \gamma^* \in \{0,1\}\}$ containing $\Phi(\frac{1}{2})$. Let $\widetilde{\mathcal{V}}$ denote the linear space generated by the columns of $\bigodot_{i=1}^{d} \mathcal{M}$. By applying (A.20), these columns are linearly independent. Hence, they constitute a basis for $\widetilde{\mathcal{V}}$. According to (c), $\widetilde{\Phi}(\frac{1}{2},\dots,\frac{1}{2})$ is contained in the span of the columns of $\bigodot_{i=1}^{d} \mathcal{M}$. Hence, it is also contained in $\widetilde{\mathcal{V}}$. For $\gamma^* \in \Gamma_{2\mathcal{I}_d}$, the relations (A.21) and (b) yield

$$\widetilde{\mathscr{A}_{\gamma^*}}\Big(\bigodot_{i=1}^{d} \mathcal{M}\Big) = \bigodot_{i=1}^{d} \mathscr{A}_{\gamma_i^*}\mathcal{M}. \tag{A.22}$$

Since the columns of $\mathscr{A}_{\gamma_i^*}\mathcal{M}$ are contained in $\mathcal{V}$, they can be spanned by the columns of $\mathcal{M}$. Thus, $\widetilde{\mathscr{A}_{\gamma^*}}\widetilde{\mathcal{V}} \subset \widetilde{\mathcal{V}}$, and $\widetilde{\mathcal{V}}$ is a common invariant subspace of $\{\widetilde{\mathscr{A}_{\gamma^*}} : \gamma^* \in \Gamma_{2\mathcal{I}_d}\}$, which contains $\widetilde{\Phi}(\frac{1}{2},\dots,\frac{1}{2})$. Since $\mathcal{V}$ is minimal, $\widetilde{\mathcal{V}}$ is also minimal. This concludes the proof.

□

Applying the relations of Lemma A.2.3 provides an alternative proof of Proposition 6.4.11. Since we only deal with the mask, it requires a reformulation of the result:

Proposition A.2.4. *Given a univariate dyadic symbol a with* $\operatorname{supp}(a_k)_{k\in\mathbb{Z}} \subset [0,L]$, *let it generate a continuous refinable function φ with globally linearly independent integer shifts. Then the d-dimensional tensor product $\widetilde{\varphi} = \bigotimes_{i=1}^{d} \varphi$ has locally linearly independent integer shifts.*

Proof. Due to Theorem 6.4.1, φ has locally linearly independent integer shifts. Note that $\widetilde{\varphi}$ is continuous, and it is refinable with respect to the symbol $\widetilde{a} := \bigotimes_{i=1}^{d} a$. In order to simplify notation, we confine ourselves to $d=2$. The general multivariate setting can be proven in a similar way, but with much complexer notation. For $l \in \mathbb{N}_0$, let

$$\gamma_1^* = \begin{pmatrix}\gamma_{1,1}^*\\ \gamma_{1,2}^*\end{pmatrix}, \dots, \gamma_l^* = \begin{pmatrix}\gamma_{l,1}^*\\ \gamma_{l,2}^*\end{pmatrix} \in \{0,1\}^2.$$

Then we have

$$\begin{aligned}\widetilde{\mathscr{A}_{\gamma_1^*}} \cdots \widetilde{\mathscr{A}_{\gamma_l^*}} \widetilde{\mathcal{M}} &= \left(\mathscr{A}_{\gamma_{1,1}^*} \otimes \mathscr{A}_{\gamma_{1,2}^*}\right) \cdots \left(\mathscr{A}_{\gamma_{l,1}^*} \otimes \mathscr{A}_{\gamma_{l,2}^*}\right)(\mathcal{M}\otimes\mathcal{M})\\ &= \left(\mathscr{A}_{\gamma_{1,1}^*}\cdots\mathscr{A}_{\gamma_{l,1}^*}\mathcal{M}\right) \otimes \left(\mathscr{A}_{\gamma_{1,2}^*}\cdots\mathscr{A}_{\gamma_{l,2}^*}\mathcal{M}\right).\end{aligned}$$

Let us use the short-hand notation

$$\widetilde{\mathscr{L}} := \widetilde{\mathscr{A}_{\gamma_1^*}} \cdots \widetilde{\mathscr{A}_{\gamma_l^*}} \widetilde{\mathcal{M}}$$

as well as

$$\mathscr{L}_1 := \mathscr{A}_{\gamma_{1,1}^*}\cdots\mathscr{A}_{\gamma_{l,1}^*}\mathcal{M}, \qquad \mathscr{L}_2 := \mathscr{A}_{\gamma_{1,2}^*}\cdots\mathscr{A}_{\gamma_{l,2}^*}\mathcal{M}.$$

We have $\operatorname{card}(\widetilde{K}) = \operatorname{card}(K)^2$ and $\widetilde{\mathscr{L}} = \mathscr{L}_1 \otimes \mathscr{L}_2$. Let us denote by zr and nzr the number of zero and nonzero rows of a matrix, respectively. This yields

$$\begin{aligned}
\operatorname{rank}(\widetilde{\mathscr{L}}) &= \operatorname{rank}(\mathscr{L}_1)\cdot\operatorname{rank}(\mathscr{L}_2)\\
&= \big(\operatorname{card}(K)-\operatorname{zr}(\mathscr{L}_1)\big)\cdot\big(\operatorname{card}(K)-\operatorname{zr}(\mathscr{L}_2)\big)\\
&= \operatorname{card}(K)^2-\big(\operatorname{zr}(\mathscr{L}_1)\cdot\operatorname{card}(K)+\operatorname{card}(K)\cdot\operatorname{zr}(\mathscr{L}_2)-\operatorname{zr}(\mathscr{L}_1)\cdot\operatorname{zr}(\mathscr{L}_2)\big)\\
&= \operatorname{card}(K)^2-\big(\operatorname{zr}(\mathscr{L}_1)\cdot\operatorname{card}(K)+(\operatorname{card}(K)-\operatorname{zr}(\mathscr{L}_1))\cdot\operatorname{zr}(\mathscr{L}_2)\big)\\
&= \operatorname{card}(K)^2-\big(\operatorname{zr}(\mathscr{L}_1)\cdot\operatorname{card}(K)+\operatorname{nzr}(\mathscr{L}_1)\cdot\operatorname{zr}(\mathscr{L}_2)\big).
\end{aligned}$$

Since

$$\operatorname{zr}(\mathscr{L}_1)\cdot\operatorname{card}(K)+\operatorname{nzr}(\mathscr{L}_1)\cdot\operatorname{zr}(\mathscr{L}_2)$$

is precisely the number of zero rows in $\widetilde{\mathscr{L}}$, and $\operatorname{card}(\widetilde{K})=\operatorname{card}(K)^2$, we obtain

$$\operatorname{rank}(\widetilde{\mathscr{L}})=\operatorname{card}(\widetilde{K})-\operatorname{zr}(\widetilde{\mathscr{L}}).$$

Hence, the nonzero rows of $\widetilde{\mathscr{L}}$ are linearly independent. This concludes the proof. □

A.3 Auxiliary Notation and Results

Filter Bank Notation

In the sequel, we introduce some filter bank notation. The convolution of a symbol $a=\frac{1}{m}\sum_{k\in\mathbb{Z}^d}a_k e^{-2\pi ik\cdot\xi}$ with a sequence $(c_k)_{k\in\mathbb{Z}^d}$ means the ordinary convolution with its coefficient sequence. However, our normalization requires the introduction of a factor $\frac{1}{\sqrt{m}}$, i.e,

$$a*(c_k)_{k\in\mathbb{Z}^d}:=\frac{1}{\sqrt{m}}(a_k)_{k\in\mathbb{Z}^d}*(c_k)_{k\in\mathbb{Z}^d}. \tag{A.23}$$

The *downsampling operator* $\downarrow_M$ means

$$(c_k)_{k\in\mathbb{Z}^d}\downarrow_M:=(c_{Mk})_{k\in\mathbb{Z}^d},$$

and *upsampling* $\uparrow_M$ is given by

$$(c_k)_{k\in\mathbb{Z}^d}\uparrow_M:=\left(\begin{cases}c_l, & \text{for } k=Ml\\ 0, & \text{otherwise}\end{cases}\right)_{k\in\mathbb{Z}^d}.$$

Note that if we take $\downarrow_M$ and $\uparrow_M$ as operators on $\ell_2(\mathbb{Z}^d)$, then they are adjoint to each other.

The Riesz-Thorin Interpolation Theorem

The following theorem is known as the Riesz-Thorin Interpolation Theorem, see for instance the textbook [Wer97]. In order to reduce effort, we only recall the result concerning matrix operators on sequence spaces:

Theorem A.3.1. *Given $1\le p_1,p_2,q_1,q_2\le\infty$ and $0\le\theta\le1$, let U be a matrix operator, which is bounded from ℓ_{p_i} to ℓ_{q_i} with operator norm C_i, $i=1,2$. Then, for*

$$\frac{1}{p}=(1-\theta)\frac{1}{p_1}+\theta\frac{1}{p_2},\qquad \frac{1}{q}=(1-\theta)\frac{1}{q_1}+\theta\frac{1}{q_2},$$

U is bounded from ℓ_p to ℓ_q with operator norm $C_1^{1-\theta}C_2^{\theta}$.

List of Figures

List of Tables

Bibliography

[BGN04] L. Borup, R. Gribonval, and M. Nielsen, *Bi-framelet systems with few vanishing moments characterize Besov spaces*, Appl. Comput. Harm. Anal. **17** (2004), no. 1, 3–28.

[BN] L. Borup and M. Nielsen, *Some remarks on shrinkage operators*, preprint.

[Bow00] M. Bownik, *A characterization of affine dual frames in $L_2(\mathbb{R}^d)$*, Appl. Comput. Harm. Anal. **8** (2000), 203–221.

[CD93] A. Cohen and I. Daubechies, *Non-separable bi-dimensional wavelet bases*, Rev. Mat. Iberoamericana **9** (1993), 51–137.

[CDD00] A. Cohen, W. Dahmen, and R. A. DeVore, *Multiscale decompositions on bounded domains*, Trans. Amer. Math. Soc. **352** (2000), no. 8, 3651–3685.

[CDF92] A. Cohen, I. Daubechies, and J.-C. Feauveau, *Biorthogonal bases of compactly supported wavelets*, Comm. Pure Appl. Math. **45** (1992), 485–500.

[CDH00] A. Cohen, R. A. DeVore, and R. Hochmuth, *Restricted nonlinear approximation*, Constr. Approx. **16** (2000), 85–113.

[CDLL98] A. Chambolle, R. A. DeVore, N. Y. Lee, and B. J. Lucier, *Nonlinear wavelet image processing: variational problems, compression, and noise removal through wavelet shrinkage*, IEEE Trans. Image Process. **7** (1998), 319–335.

[CDM91] A. S. Cavaretta, W. Dahmen, and C. A. Micchelli, *Stationary subdivision*, Mem. Amer. Math. Soc. **93** (1991), 1–186.

[CDPX99] A. Cohen, R. DeVore, P. Petrushev, and H. Xu, *Nonlinear approximation and the space $BV(\mathbb{R}^2)$*, Amer. J. Math. **121** (1999), 587–628.

[CH01] C. K. Chui and W. He, *Construction of multivariate tight frames via kronecker products*, Appl. Comput. Harm. Anal. **11** (2001), 305–312.

[CHR00] D. R. Chen, B. Han, and S. D. Riemenschneider, *Construction of multivariate biorthogonal wavelets with arbitrary vanishing moments*, Adv. Comput. Math. **13** (2000), no. 2, 131–165.

[Chr03] O. Christensen, *An Introduction to Frames and Riesz Bases*, Birkhäuser, Boston, 2003.

[CHS02] C. K. Chui, W. He, and J. Stöckler, *Compactly supported tight and sibling frames with maximum vanishing moments*, Appl. Comput. Harm. Anal. **13** (2002), 224–262.

[CL94] C. K. Chui and C. Li, *A general framework of multivariate wavelets with duals*, Appl. Comput. Harm. Anal. **1** (1994), 368–390.

[Coh03] A. Cohen, *Numerical Analysis of Wavelet Methods*, vol. 32, Studies in Mathematics and its Applications, Amsterdam, 2003.

[CSS98] C. K. Chui, X. Shi, and J. Stöckler, *Affine frames, quasi-affine frames and their duals*, Adv. Comput. Math. **8** (1998), 1–17.

[CT97] C. Canuto and A. Tabacco, *Multilevel decompositions of functional spaces*, J. Fourier Anal. Appl. **3** (1997), no. 6, 715–742.

[Dau88] I. Daubechies, *Orthonormal bases of compactly supported wavelets*, Comm. Pure Appl. Math. **41** (1988), 909–996.

[Dau92] ———, *Ten Lectures on Wavelets*, vol. 9, SIAM, Philadelphia, 1992, CBMS-NSF Regional Conf. Ser. in Appl. Math. 61.

[dBDR93] C. de Boor, R. A. DeVore, and A. Ron, *On the construction of multivariate (pre)wavelets*, Constr. Approx. **9** (1993), 123–166.

[dBHR93] C. de Boor, K. Hölling, and S. D. Riemenschneider, *Box Splines*, Springer, New York, 1993.

[DD97] S. Dahlke and R. A. DeVore, *Besov regularity for elliptic boundary value problems*, Commun. Partial Diff. Eqns. **22** (1997), 1–16.

[DDD97] S. Dahlke, W. Dahmen, and R. A. DeVore, *Nonlinear approximation and adaptive techniques for solving elliptic operator equations*, Multiscale Wavelet Methods for Partial Differential Equations (W. Dahmen, A. Kurdila, and P. Oswald, eds.), 1997, pp. 237–283.

[DDD04] I. Daubechies, M. Defrise, and C. DeMol, *An iterative thresholding algorithm for linear inverse problems with a sparsity constraint*, Comm. Pure Appl. Math. **57** (2004), 1413–1541.

[DDU02] S. Dahlke, W. Dahmen, and K. Urban, *Adaptive wavelet methods for saddle point problems – optimal convergence rates*, SIAM J. Numer. Anal. **40** (2002), no. 4, 1230–1262.

[Der99] J. Derado, *Multivariate refinable interpolating functions*, Appl. Comput. Harm. Anal. **7** (1999), 165–183.

[DeV98] R. A. DeVore, *Nonlinear approximation*, Acta Numerica (1998), 51–150.

[DGM99] S. Dahlke, K. Gröchenig, and P. Maass, *A new approach to interpolating scaling functions*, Appl. Anal. **72** (1999), 485–500.

[DH00] I. Daubechies and B. Han, *Pairs of dual wavelet frames from any two refinable functions*, Constr. Approx. **20** (2000), no. 3, 325–352.

[DHRS03] I. Daubechies, B. Han, A. Ron, and Z. Shen, *Framelets: MRA-based constructions of wavelet frames*, Appl. Comput. Harm. Anal. **14** (2003), no. 1, 1–46.

[DJ94] D. Donoho and I. M. Johnstone, *Ideal spatial adaptation by wavelet shrinkage*, Biometrika **81** (1994), 425–455.

[DJP92] R. A. DeVore, B. Jawerth, and V. A. Popov, *Compression of wavelet decompositions*, American Journal of Math. **114** (1992), 737–785.

[DL93] R. A. DeVore and G. G. Lorentz, *Constructive Approximation*, Springer-Verlag, 1993.

[DL04] S. Dekel and D. Leviatan, *The Bramble-Hilbert lemma for convex domains*, SIAM J. Math. Anal. **35** (2004), 1203–1212.

[DM97] S. Dahlke and P. Maass, *Interpolating refinable functions and wavelets for general scaling matrices*, Numer. Funct. Anal. Optim. **18** (1997), no. 5&6, 521–539.

[DS93] R. A. DeVore and R. Sharpley, *Besov spaces on domains in* $\mathbb{R}^d$, Trans. Amer. Math. Soc. **335** (1993), 843–864.

[DS98] I. Daubechies and W. Sweldens, *Factoring wavelet transforms into lifting steps*, J. Fourier Anal. Appl. **4** (1998), no. 3, 245–267.

[DT96] R. A. DeVore and V. N. Temlyakov, *Some remarks on greedy algorithms*, Adv. Comput. Math. **5** (1996), no. 2–3, 173–187.

[Duf56] C. C. Duffee, *The Theory of Matrices*, Chelsea Publishing Company, New York, 1956.

[Ehl] M. Ehler, *Compactly supported multivariate pairs of dual wavelet frames obtained by convolution*, to appear in: Int. J. Wavelets Multiresolut. Inf. Process.

[Ehl07] ———, *On multivariate compactly supported bi-frames*, J. Fourier Anal. Appl. **13** (2007), no. 5, 511–532.

[FS72] C. Fefferman and E. M. Stein, H^p *spaces of several variables*, Acta Math. **129** (1972), 137–193.

[FS03] H. G. Feichtinger and T. Strohmer, *Advances in Gabor Analysis*, Birkhäuser, Boston, 2003.

[Gao98] H. Y. Gao, *Wavelet shrinkage denoising using the non-negative garotte*, J. Comput. Graph. Statist. **7** (1998), no. 4, 469–488.

[GC04] K. Gröchenig and E. Cordero, *Localization of frames II*, Appl. Comp. Harm. Anal. **17** (2004), 29–47.

[GF05] K. Gröchenig and M. Fornasier, *Intrinsic localization of frames*, Constr. Approx. **22** (2005), 395–415.

[GHT04] G. Garrigos, R. Hochmuth, and A. Tabacco, *Wavelet characterizations for anisotropic Besov spaces with* $0 < p < 1$, Proc. of the Edinburgh Math. Soc. **47** (2004), 573–595.

[GM92] K. Gröchenig and W. R. Madych, *Multiresolution analysis, Haar basis, and self-similar tilings of* $\mathbb{R}^d$, IEEE Trans. Inform. Theory **38** (1992), 556–568.

[GN04] R. Gribonval and M. Nielsen, *Nonlinear approximation with dictionaries. I. Direct estimates*, J. Fourier Anal. Appl. **10** (2004), no. 1, 51–71.

[GR98] K. Gröchenig and A. Ron, *Tight compactly supported wavelet frames of arbitrarily high smoothness*, Proc. Amer. Math. Soc. **126** (1998), no. 4, 1101–1107.

[Grö01] K. Gröchenig, *Foundations of Time-Frequency Analysis*, Birkhäuser, Boston, 2001.

[Grö03] ———, *Localized frames are finite unions of riesz sequences*, Adv. Comp. Math. **18** (2003), 149–157.

[Grö04] ———, *Localization of frames, banach frames, and the invertibility of the frame operator*, J. Fourier Anal. Appl. **10** (2004), no. 2, 105–132.

[Haa10] A. Haar, *Zur Theorie der orthogonalen Funktionensysteme*, Math. Annalen **69** (1910), 331–371.

[Han99] B. Han, *Analysis and construction of optimal multivariate biorthogonal wavelets with compact support*, SIAM J. Math. Anal. **31** (1999), no. 2, 274–304.

[Han02] ———, *Symmetry property and construction of wavelets with a general dilation matrix*, Linear Algebra Appl. **353** (2002), 207–225.

[Han03a] ———, *Compactly supported tight wavelet frames and orthonormal wavelets of exponential decay with a general dilation matrix*, J. Comput. Appl. Math. **155** (2003), no. 1, 43–67.

[Han03b] ———, *Multiwavelet frames from refinable function vectors*, Adv. Comput. Math. **18** (2003), 211–245.

[Han04] ———, *Symmetric multivariate orthogonal refinable functions*, Appl. Comput. Harm. Anal. **17** (2004), 277–292.

[HJ98] B. Han and R. Q. Jia, *Optimal interpolatory subdivision schemes in multidimensional spaces*, SIAM J. Numer. Anal. **36** (1998), no. 1, 105–124.

[HJ02] ———, *Quincunx fundamental refinable functions and quincunx biorthogonal wavelets*, Math. Comp. **71** (2002), no. 237, 165–196.

[Hoc02] R. Hochmuth, *Wavelet characterizations for anisotropic Besov spaces*, Appl. Comput. Harm. Anal. **12** (2002), 179–208.

[HP05] K. Hirakawa and T. Parks, *Image denoising for signal-dependent noise*, IEEE ICASSP **2** (2005), 29–32.

[HR02] B. Han and S. D. Riemenschneider, *Interpolatory biorthogonal wavelets and cbc algorithm*, Wavelet analysis and applications, AMS/IP Studies in Advanced Mathematics, vol. 25, Amer. Math. Soc., Providence, RI, 2002, pp. 119–138.

[Jaw77] B. Jawerth, *Some observations on Besov and Lizorkin-Triebel spaces*, Math. Scand. **40** (1977), 94–104.

[Jia85] R. Q. Jia, *Local linear independence of the translates of a box spline*, Constr. Approx. **1** (1985), 175–182.

[Jia93] ______, *A bernstein-type inequality associated with wavelet decomposition*, Constr. Approx **9** (1993), 299–318.

[Jia98] ______, *Approximation properties of multivariate wavelets*, Math. Comp. **67** (1998), 647–665.

[Jia99] ______, *Characterization of smoothness of multivariate refinable functions in sobolev spaces*, Trans. Amer. Math. Soc. **351** (1999), no. 10, 4089–4112.

[JM90] R. Q. Jia and C. A. Micchelli, *Using the refinement equation for the construction of pre-wavelets II: powers of two*, Papers from the International Conference on Curves and Surfaces (Pierre-Jean Laurent and Schumaker, eds.), 1990, pp. 209–246.

[JRS99] H. Ji, S. D. Riemenschneider, and Z. Shen, *Multivariate compactly supported fundamental refinable functions, duals and biorthogonal wavelets*, Stud. Appl. Math. **102** (1999), 173–204.

[Koc] K. Koch, *Multivariate symmetric interpolating scaling vectors with duals*, to appear in: J. Fourier Anal. Appl.

[Koc07] ______, *Multivariate orthonormal interpolating scaling vectors*, Appl. Comput. Harm. Anal. **22** (2007), no. 2, 198–216.

[KS00] J. Kovačević and W. Sweldens, *Wavelet families of increasing order in arbitrary dimensions*, IEEE Trans. Image Proc. **9** (2000), no. 3, 480–496.

[Kyr96] G. Kyriazis, *Wavelet coefficients measuring smoothness in $H_p(\mathbb{R}^d)$*, Appl. Comput. Harmon. Anal. **3** (1996), no. 2, 100–119.

[Kyr01] ______, *Non-linear approximation and interpolation spaces*, J. Approx. **113** (2001), 110–126.

[Lin05] M. Lindemann, *Approximation properties of non-separable wavelet bases with isotropic scaling matrices and their relation to besov spaces*, Ph.D. thesis, University of Bremen, 2005.

[LLS97] W. Lawton, S. L. Lee, and Z. Shen, *Stability and orthonormality of multivariate refinable functions*, SIAM J. Math. Anal. **28** (1997), no. 4, 999–1014.

[Lor07] D. Lorenz, *Non-convex variational denoising of images: interpolation between hard and soft wavelet shrinkage*, Current Development in Theory and Applications of Wavelets **1** (2007), no. 1, 31–56.

[LS] M. J. Lai and J. Stöckler, *Construction of compactly supported tight wavelet frames*, to appear in: Appl. Comput. Harm. Anal.

[Mal89] S. Mallat, *Multiresolution approximation and wavelet orthonormal bases of* $L_2(\mathbb{R}^d)$, Trans. Amer. Math. Soc. **315** (1989), 69–87.

[Mal99] ———, *A Wavelet Tour of Signal Processing*, Academic Press, 1999.

[MC97] Y. Meyer and R. Coifman, *Wavelets. Calderon zygmund and multilinear operators*, Cambridge University Press, 1997.

[Mey90] Y. Meyer, *Ondelettes et Operateurs I: Ondelettes*, Hermann, Paris, 1990.

[Mey01] ———, *Oscillating patterns in image processing and nonlinear evolution equations*, American Mathematical Society, Boston, MA, USA, 2001, the fifteenth Dean Jacqueline B. Lewis memorial lectures.

[Mey86] ———, *Principe d'incertitude, bases hilbertienne et algebres d'operateurs*, Semin. Bourbaki **38** (1985/86), no. 662, 209–223.

[MP03] L. B. Montefusco and S. Papi, *A parameter selection method for wavelet shrinkage denoising*, BIT **43** (2003), 611–626.

[MPMK98] A. Mojsilovic, M. Popovic, S. Markovic, and M. Krstic, *Characterization of visually similar diffuse diseases from B-scan liver images using nonseparable wavelet transform*, IEEE Trans. on Med. Imaging **17** (1998), no. 4, 541–549.

[Par95] H. Park, *A computational theory of laurent polynomial rings and multidimensional fir systems*, Ph.D. thesis, UC Berkeley, 1995.

[Pee76] J. Peetre, *New Thoughts on Besov Spaces*, Duke Univ. Math. Series, Durham, 1976.

[Qui76] D. Quillen, *Projective modules over polynomial rings*, Invent. Math **36** (1976), 167–171.

[ROF92] L. Rudin, S. Osher, and E. Fatemi, *Nonlinear total variation based noise removal algorithms*, Physica D **60** (1992), 259–268.

[Ron99] A. Ron, *Introduction to shift-invariant spaces I: Linear independence*, Multivariate Approximation and Applications (N. Dyn et al., ed.), 1999, pp. 112–151.

[RS91] S. D. Riemenschneider and Z. Shen, *Box splines, cardinal series and wavelets*, Approximation Theory and Functional Analysis (C. K. Chui, ed.), Academic Press, New York, 1991, pp. 133–149.

[RS92] ———, *Wavelets and pre-wavelets in low dimensions*, J. Approx. Theory **71** (1992), 18–38.

[RS96] T. Runst and W. Sickel, *Sobolev Spaces of Fractional Order, Nemytskij Operators, and Nonlinear Partial Differential Equations*, Walter de Gruyter, Berlin, 1996.

[RS97a] S. D. Riemenschneider and Z. Shen, *Multidimensional interpolatory subdivision schemes*, SIAM J. Numer. Anal. **34** (1997), 2357–2381.

[RS97b] A. Ron and Z. Shen, *Affine systems in $L_2(\mathbb{R}^d)$ II: dual systems*, J. Fourier Anal. Appl. **3** (1997), 617–637.

[RS97c] ———, *Affine systems in $L_2(\mathbb{R}^d)$: the analysis of the analysis operator*, J. Funct. Anal. **148** (1997), 408–447.

[RS98] ———, *Compactly supported tight affine spline frames in $L_2(\mathbb{R}^d)$*, Math. Comp. **67** (1998), 191–207.

[Ryc99] V. Rychkov, *On restrictions and extensions of the Besov and Triebel-Lizorkin spaces with respect to Lipschitz domains*, J. London Math. Soc. **60** (1999), no. 1, 237–257.

[SA04] I. W. Selesnick and A. F. Abdelnour, *Symmetric wavelet tight frames with two generators*, Appl. Comput. Harm. Anal. **17** (2004), no. 2, 211–225.

[SN96] G. Strang and T. Nguyen, *Wavelets and Filter Banks*, Wellesley-Cambridge Press, Wellesley, 1996.

[Str81] J.-O. Strömberg, *A modified Franklin system and higher-order spline systems on $\mathbb{R}^d$ as unconditional bases for Hardy spaces*, Conf. on Harmonic Analysis in Honor of A. Zygmund (1981), 475–494.

[Sus76] A. Suslin, *Projective modules over polynomial rings are free*, Soviet Math. Dokl. **17** (1976), 1160–1164.

[Swa78] R. G. Swan, *Projective modules over laurent polynomial rings*, Trans. Amer. Math. Soc. **237** (1978), 111–120.

[Swe96] W. Sweldens, *The lifting scheme: a custom-design construction of biorthogonal wavelets*, Appl. Comput. Harm. Anal. **3** (1996), no. 2, 186–200.

[Swe97] W. Sweldens, *The lifting scheme: A construction of second generation wavelets*, SIAM J. Math. Anal. **29** (1997), no. 2, 511–546.

[Tao96] T. Tao, *On the almost everywhere convergence of wavelet summation methods*, Appl. Comput. Harm. Anal. **3** (1996), 384–387.

[Tem98] V. N. Temlyakov, *The best m-term approximation and greedy algorithms*, Adv. Comput. Math. **8** (1998), no. 3, 249–265.

[TFG01] H. Tian, B. Fowler, and A. Gamal, *Analysis of temporal noise in CMOS photodiode active pixel sensor*, IEEE Journal of Solid-State Circuits **36** (2001), no. 1, 92–101.

[Trè67] F. Trèves, *Topological Vector Spaces, Distributions, and Kernels*, Acad. Press, New York, 1967.

[Tri83] H. Triebel, *Theory of Function Spaces*, Birkhäuser, Basel, 1983.

[Tri92] ———, *Theory of Function Spaces II*, Birkhäuser, Basel, 1992.

[Vil94] L. Villemoes, *Wavelet analysis of refinement equations*, SIAM J. Math. Anal. **25** (1994), 1433–1460.

[VSo] http://www.mathematik.uni-marburg.de/~dahlke/ag-numerik/research/software/.

[Wer97] D. Werner, *Funktionalanalysis*, Springer, Berlin, 1997.

[Woj97] P. Wojtaszczyk, *A Mathematical Introduction to Wavelets*, Cambridge University Press, 1997.